IMAGES
of America

The Anaconda Copper Mining Company of Great Falls

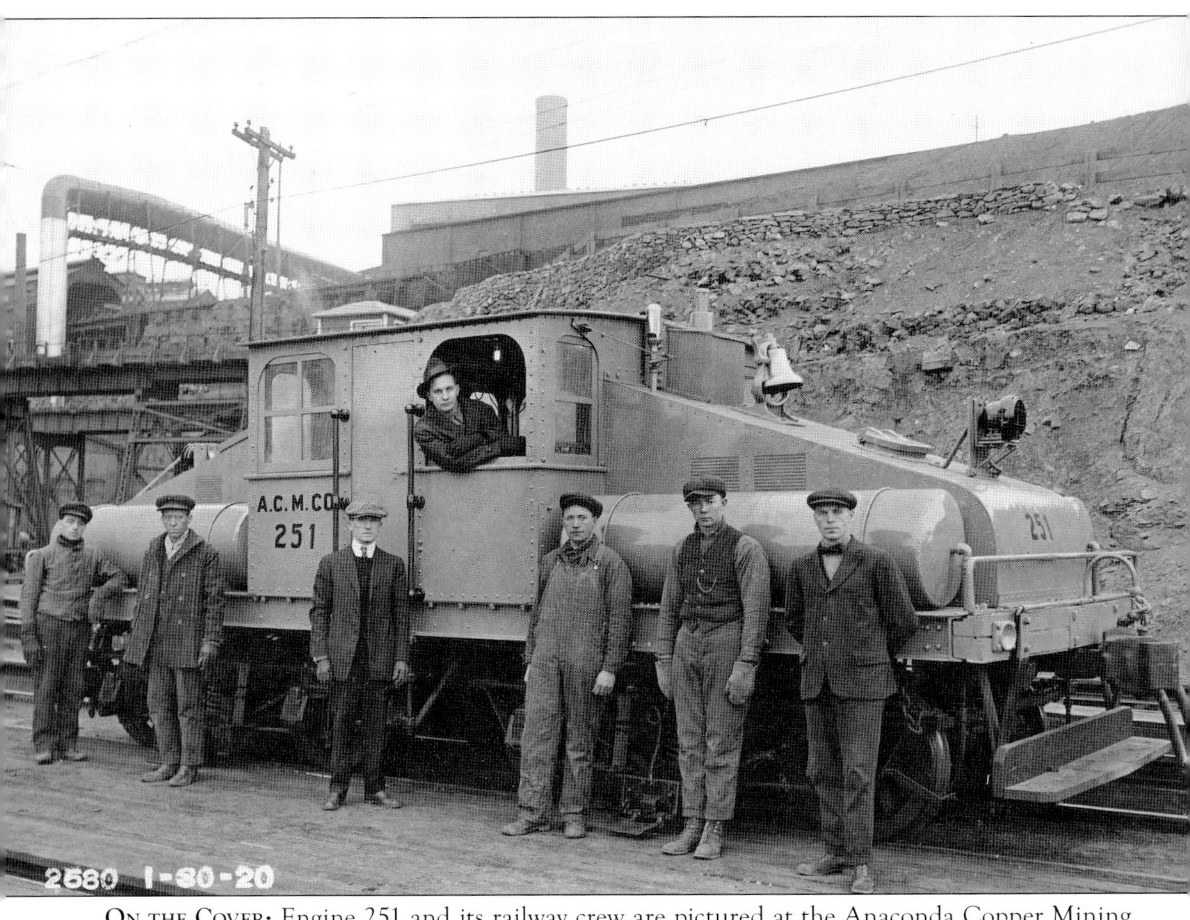

On the Cover: Engine 251 and its railway crew are pictured at the Anaconda Copper Mining Company Plant in Great Falls, Montana, on January 30, 1920. (Courtesy of the Cascade Historical Society 1982.017.0290.)

IMAGES
of America

The Anaconda Copper Mining Company of Great Falls

Megan Sanford

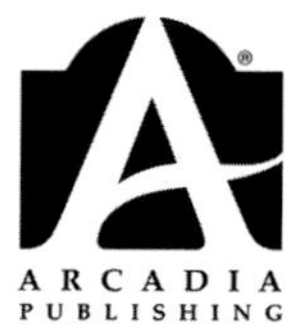

ISBN 978-1-4671-6333-0

Published by Arcadia Publishing
Charleston, South Carolina

Printed in the United States of America

Library of Congress Control Number: 2025949132

For all general information, please contact Arcadia Publishing:
Telephone 843-853-2070
Fax 843-853-0044
E-mail sales@arcadiapublishing.com

Visit us on the Internet at www.arcadiapublishing.com

To Mom, Dad, and my husband, Jon, for letting me do what I love, and to Dr. Vernon Pedersen for being my guide.

CONTENTS

Acknowledgments

The Cascade County Historical Society's History Museum & Research Center of Great Falls, Montana, is the sole source of photographs and information for this book. All images are courtesy of its collection. Even the newspaper articles and books that a couple of the images came from are housed at the research center. It has been able to protect and share these images and information because of the pioneers of Cascade County and their descendants. None of this would be possible without them. Each photograph has an accompanying number that identifies it within the collections. For researchers and those who have an interest in the Great Falls Reduction Department Plant, the numbers will make it easier to assist at the research center.

I would like to acknowledge the previous executive director, Kristi Scott, for allowing me to pursue this project. The collections curator, Ashleigh McCann, diligently took photographs of the objects in our collections so that they could also be shared. The board of directors and new executive director Connie Constan were supportive of wanting to share the resources in the archives and collections with a wider audience.

I would also like to acknowledge the communities of Black Eagle and Great Falls, which were the homes of the smelter workers. They supported the workers, their families, and their livelihoods. I hope this book will serve as a fitting tribute.

Introduction

When the call of gold came from California, the Montana Territory was bypassed, its treasures unknown and unrecognized. It was a vast land, still on the edge of lawlessness and freedom, considered only fit for fur traders and Native Americans. As the gold petered out on the West Coast, however, those still thirsty to strike it rich worked their way back east into the Montana wilds. Gold was found at Bannack, Alder Gulch, Last Chance Gulch, and many other sites, bringing the placer miners in droves. The birth of the mining industry in the West came from the lust of gold and silver. What became the Butte Mining District and many other districts in the territory were staked as silver claims.

Two men, William Allison and G.O. Humphreys, were the first to stake a claim in the area that would become the city of Butte, Montana. They began their mining in what was called the Original Lode claim in 1864. While placer miners came and went, Humphreys and Allison developed their mine and opened new ones still in search of gold and silver. They had found copper at this point, but it was not commercially relevant yet. Copper was used only for sheathing ships, roofing buildings, and cooking utensils at the time. It was more considered a nuisance, keeping them from their desired treasures. There was also no plant for processing copper; only facilities for processing silver existed in the West.

The townsite of Butte was drawn in 1867, giving organization to the mining camp. The 1870s were the peak of Butte's silver era and made the town a permanent fixture on the map. It was the leading silver-producing mining district in the world. Butte gained the nickname "the Richest Hill in the World." This was also the time that one of the main characters in Montana's history arrived: Marcus Daly. Daly was an affable Irishman who was generous and likable. He arrived in Butte with the intent of purchasing a silver mining claim for his employers, the Walker brothers of Salt Lake City, Utah. He was a mining engineer who had learned mining techniques in Nevada.

The copper era began in 1882 with the creation of Thomas Edison's Pearl Street power station in New York. Electricity gave copper a practical use, and that use expanded with communication and transportation as well. Copper is a nonferrous metal that can be found in a free metallic state or in ore. It was used by humanity long before iron by every civilization it was near. Copper had no legacy of bloodshed, though it was indispensable. But as the world became more industrious, copper was left behind.

The concentration of copper mining and processing was in the Lake Superior region of Michigan at this time. Though some small deposits had been found elsewhere, there was no reason to develop those sites, nor were they large enough to be significant. The plants in Michigan supplied the small amount that was needed in the country until the dawn of electricity. Another key figure in Montana's history was W.A. Clark. Clark and Daly were friends to begin with, both entrenched in the mining industry, but as Daly's business grew, Clark began to see him as a rival. Marcus Daly purchased what became the Anaconda Mine in 1880, expanding his empire. Clark wanted to run Butte, and he did not want to share. This led to what was known as the War of the Copper Kings.

Clark shipped some of his ore to the Boston and Colorado Smelter in Blackhawk, Colorado, but the reduction cost was too high. Clark also sent some silver ore to the same smelter, and this led to the creation of the Colorado & Montana Smelting Company, which erected a reduction works in Butte. As more silver and copper were found, processing for them both had to be developed. Smelters were built in the early 1880s, and though most processed copper, several could also treat gold and silver ore. Local processing cut production costs, but the product generated still had to be shipped east for further refining. Butte's development brought about Montana's domination in the copper mining industry. Before 1882, the Lake Superior area provided 80 percent of the United States' copper. By 1898, Butte provided 41 percent of the world's copper.

While Clark fought with Daly on a political and legal level, Daly worked with Hearst, Haggin, Tevis, and Company to create a financial foundation for his mining venture, the Anaconda Copper Mining Company. The Anaconda Copper Mining Company launched in 1895. Daly surrounded himself with mining and metallurgical engineers, lawyers, accountants, bankers, and anyone else who could help build his enterprise. Electricity and copper development went hand in hand, as one brought about change for the other. Electrolytic copper refining was pioneered, making refining copper a more lucrative endeavor and producing purer copper in large quantities.

Montana was a land of great potential for those who knew how to seize it. Industrial giants like Clark and Daly were in their element among others like James J. Hill, the railroad magnate; Conrad Kohrs, the cattle baron; and Andrew B. Hammond, the timber giant. These names are known for shaping Montana's early development. There were other successful men at city level all throughout the state as well. One of those men was Paris Gibson.

Paris Gibson was born in Maine and had made his way to Minneapolis, where he ran the Cataract Flour Mill and was quite successful. Unfortunately, the Panic of 1873 dashed his fortunes to nothing. He, like many others, saw Montana as a new beginning. He had read about the great falls of the Missouri River in the Lewis and Clark journals and wanted to see them for himself. He and his son Theodore came to Montana in 1879 with plans of establishing a sheep ranch. In 1882, he came south from Fort Benton to get a view of the falls. Gibson wrote about this trip in his memoirs: "Although I had traveled much over Northern Montana and the country between the Missouri River and the Yellowstone during my three-year residence in Fort Benton, I had never seen a spot as attractive as this and one that at once appealed to me as an ideal site for a city."

One

In the Beginning

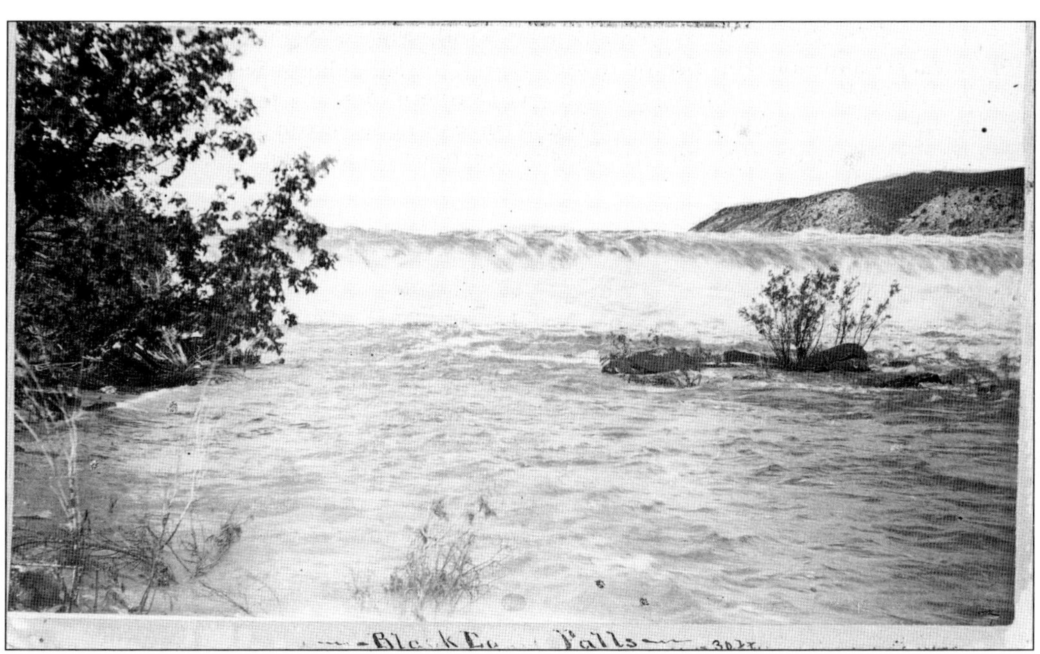

When Paris Gibson, founder of Great Falls, first saw the falls of the Missouri River, he knew that he could build a great, industrial city there. Hydroelectric power was the next big thing, and he could see the possibilities. The appeal of abundant, cheap electricity brought the Boston and Montana Consolidated Copper and Silver Mining Company to town. Above is Black Eagle Falls before it was dammed in 1890. (1983.109.0003.)

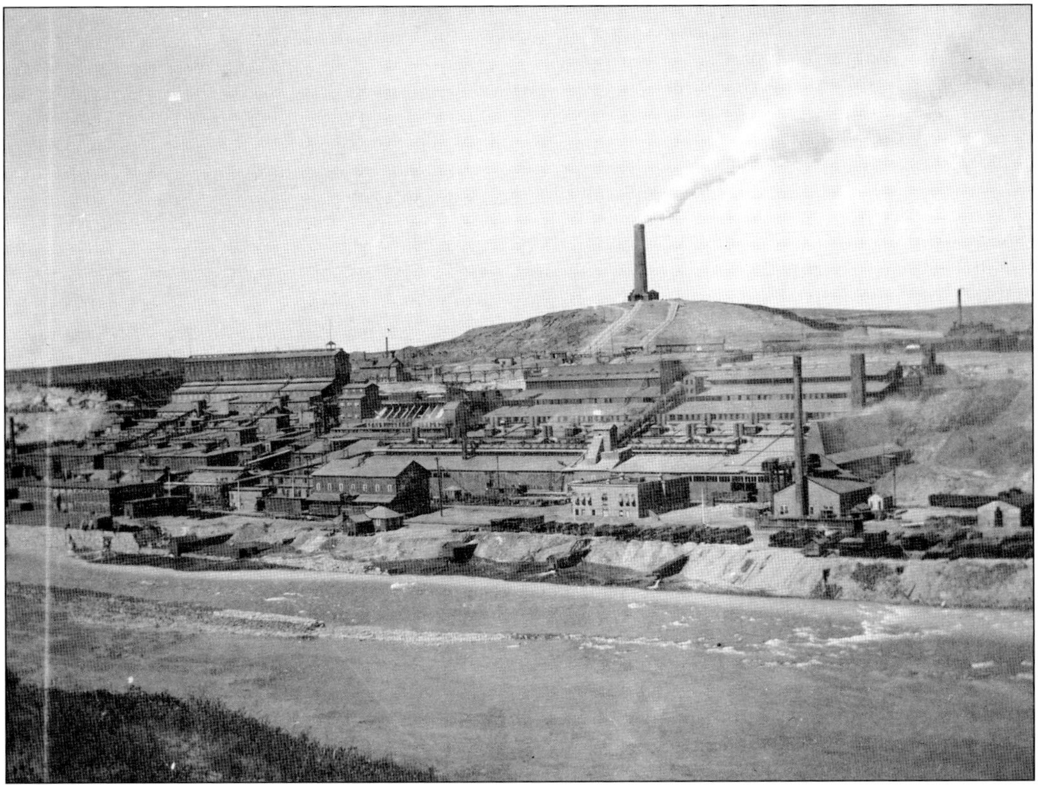

The Boston and Montana Company broke ground in 1891 and built a copper reduction plant to process the ore that came from its mines in Butte, Montana. The next year, a concentrator was installed, then roasting furnaces, reverberatory smelting furnaces, converters, and finally blast furnaces. Next, an electrolytic copper refinery and a furnace refinery were added and made the plant able to process from ore to finished copper. A powerhouse next to the Black Eagle Dam generated all the power the plant needed. (Above, 1991.114.0006; below, 1992.066.0045.)

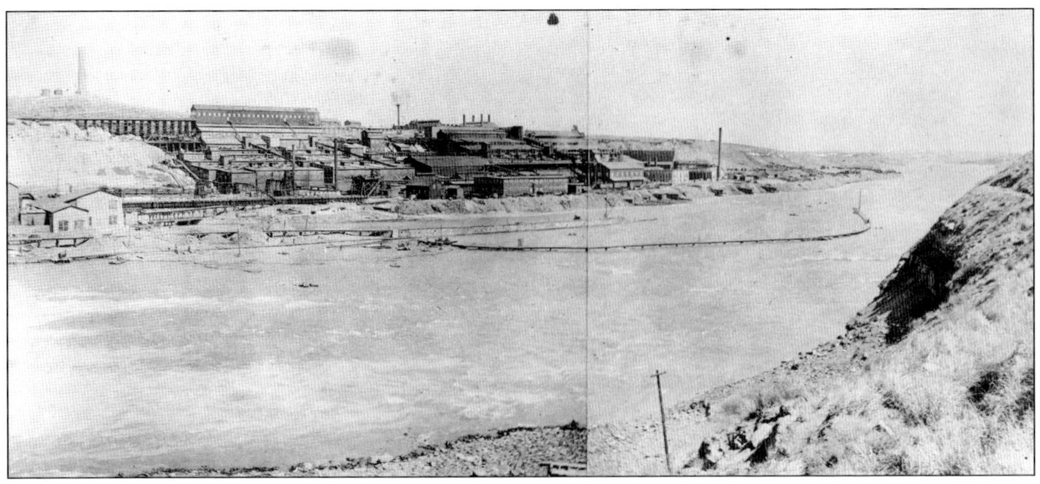

The Boston and Montana Smelter was the first in Montana to use natural gas in smelting copper. The furnaces used a double flue system that carried the gases and fumes out a smokestack. The flue was 1,600 feet long, and the stack was 186 feet tall. At right and below are photographs of the stack being built. The plant employed 900 people, not only in the refinery but also in the electric power plant and substations, offices, laboratory, and rail lines. It was the beginning of the industrial era for Great Falls. (Right, 1991.114.0006; below, 2010.027.0001.)

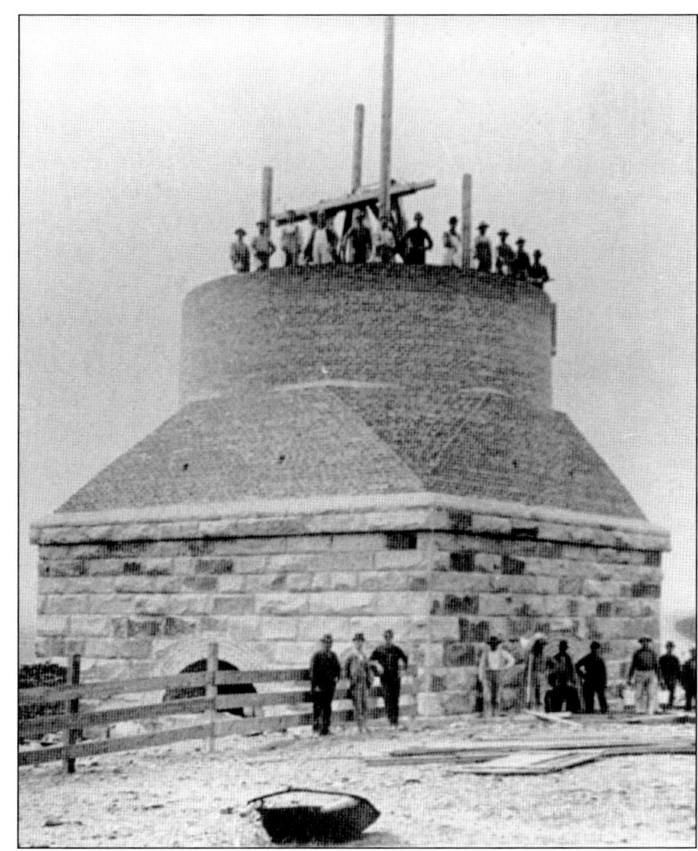

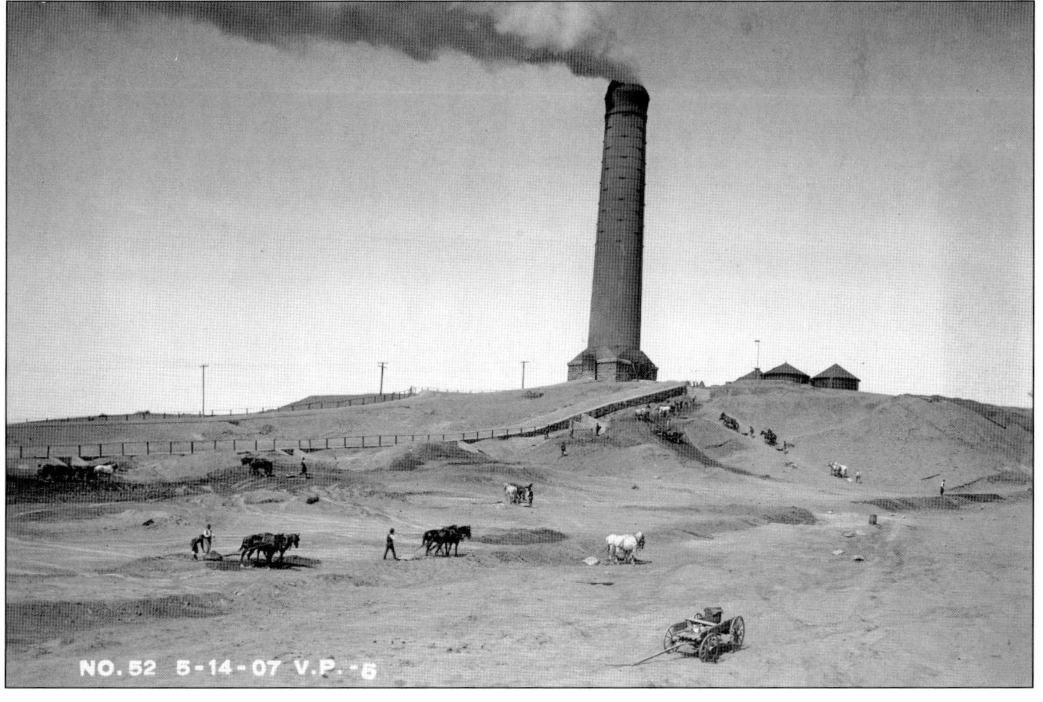

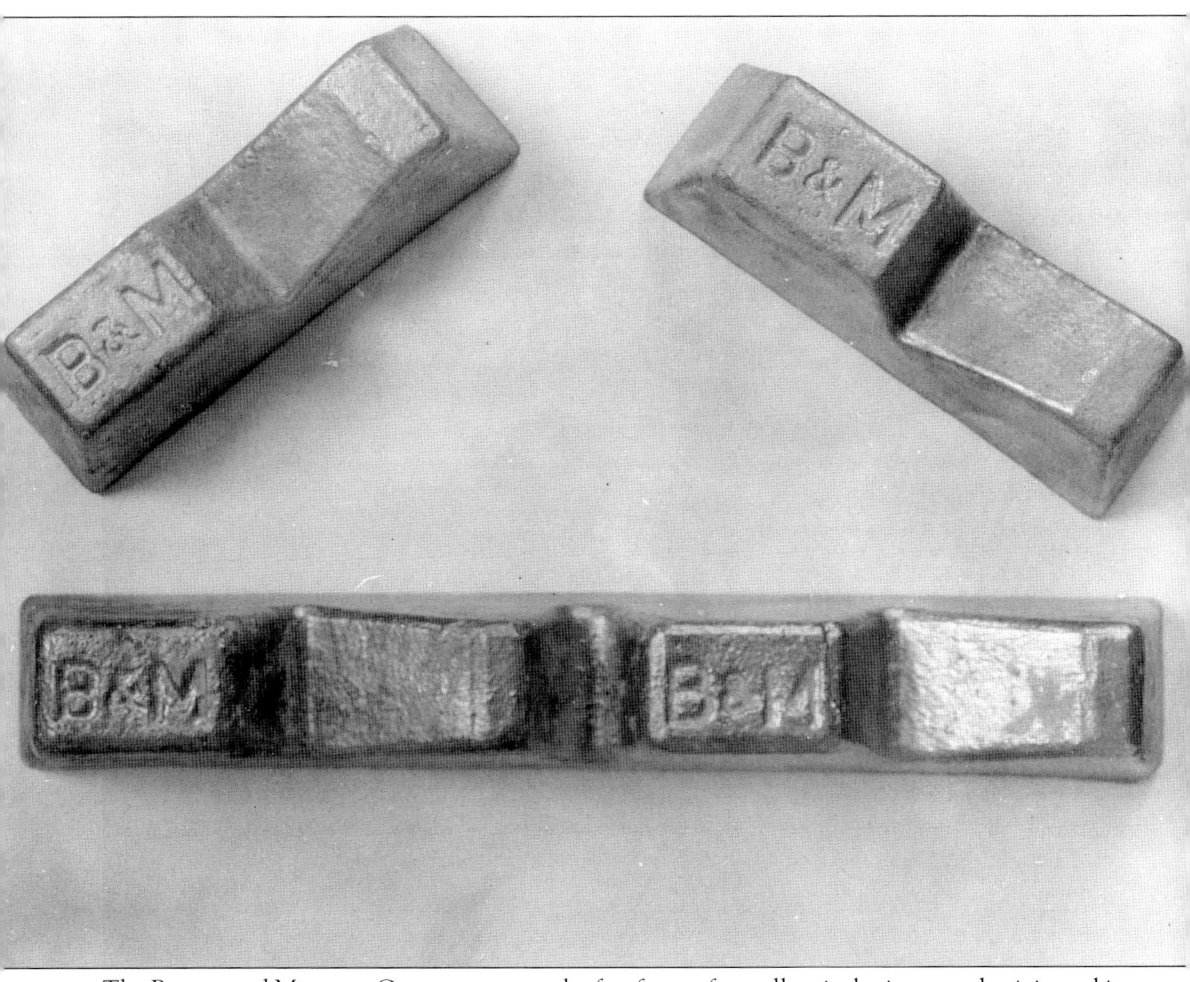

The Boston and Montana Company was at the forefront of metallurgical science and originated its own processes for refining copper. The reduction works were unique, as no other copper company had done it on such a large scale before. Above is a photograph of some of the stamped copper ingots from the Boston and Montana Smelter. (1982.017.0084.)

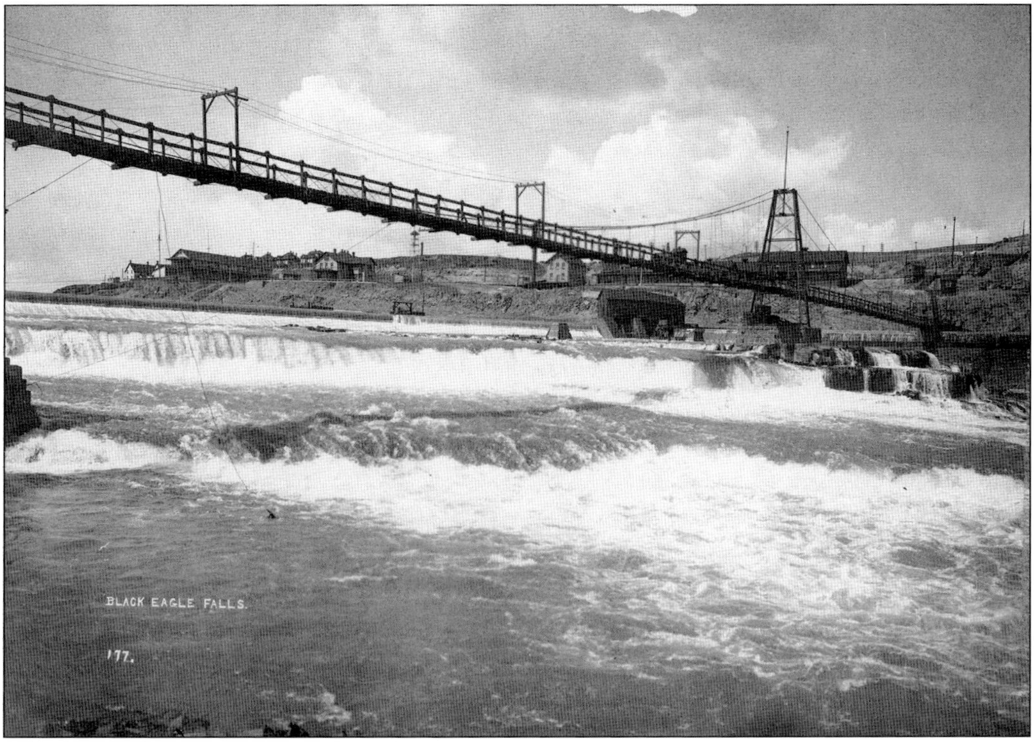

In 1891, a wagon bridge was built across the river at Fifteenth Street, but employees still faced at least a one-mile walk each way to work as only the department managers and higher ups of the plant were provided housing on smelter property. A solution was to build a wooden pedestrian bridge where the power plant was situated above Black Eagle Dam, as seen in these photographs. (Above, 1990.026.0268; below, 1999.069.0028.)

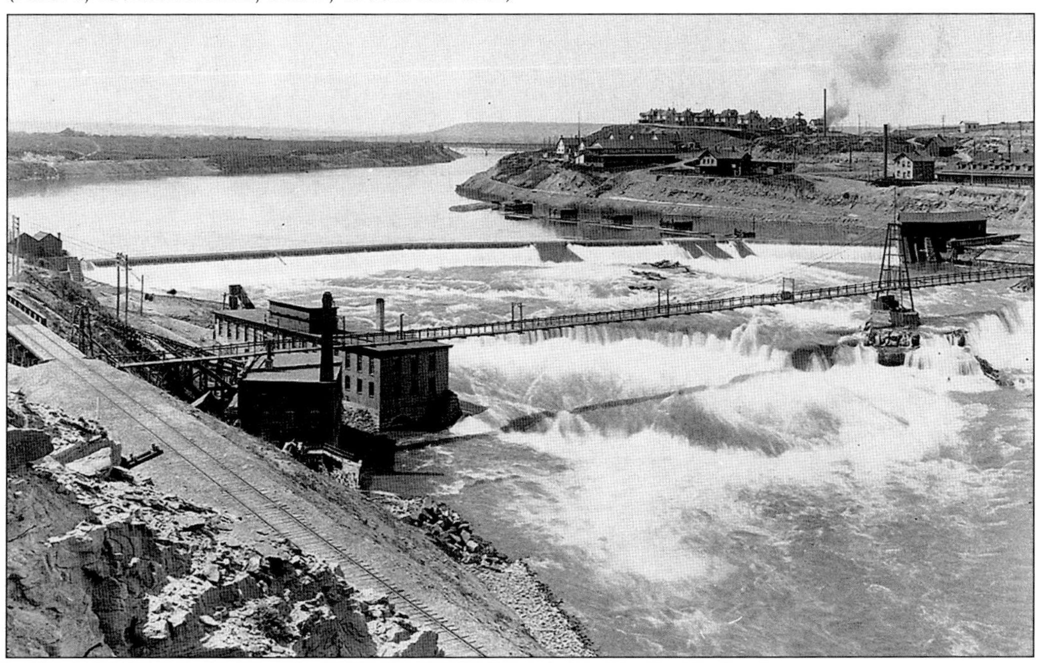

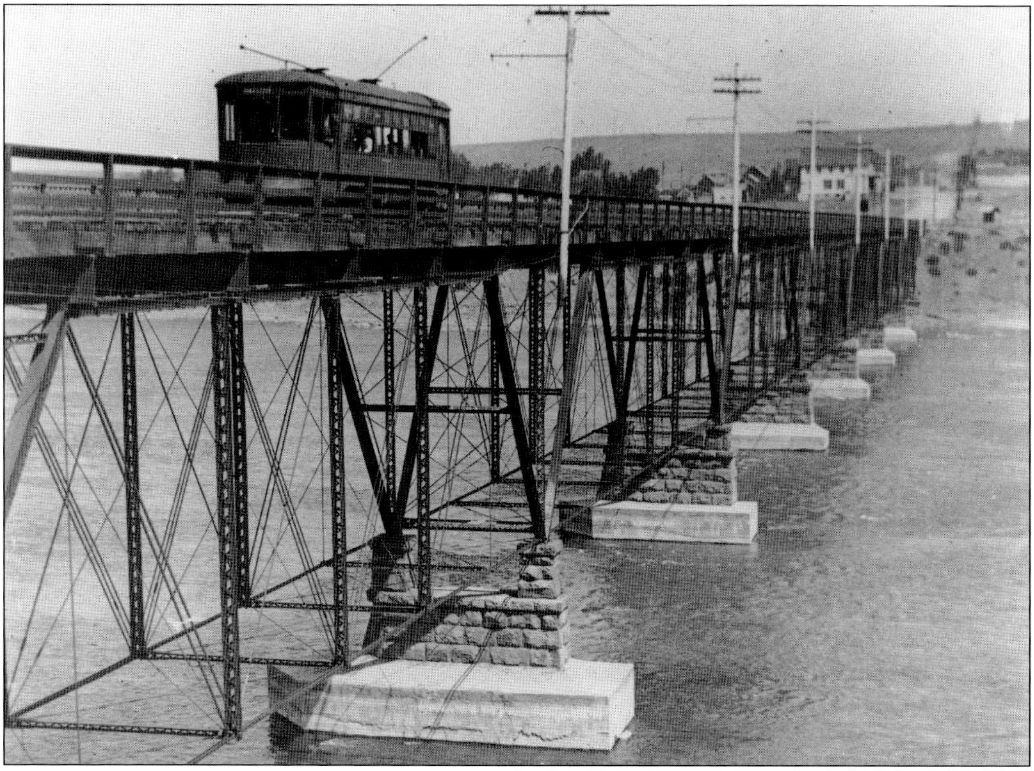

In 1900, the Great Falls Street Railway Co. extended its streetcar line and dedicated a specific car to taking workers to the smelter. Above is a streetcar crossing the Fifteenth Street bridge. The streetcars lasted until 1931, when Montana Power decided to abandon the street railway company. The City of Great Falls began a bus service the next day. Below is one of the cars that used to carry workers to the plant. (Above, 1998.072.0008; below, CCHS 133.0001.)

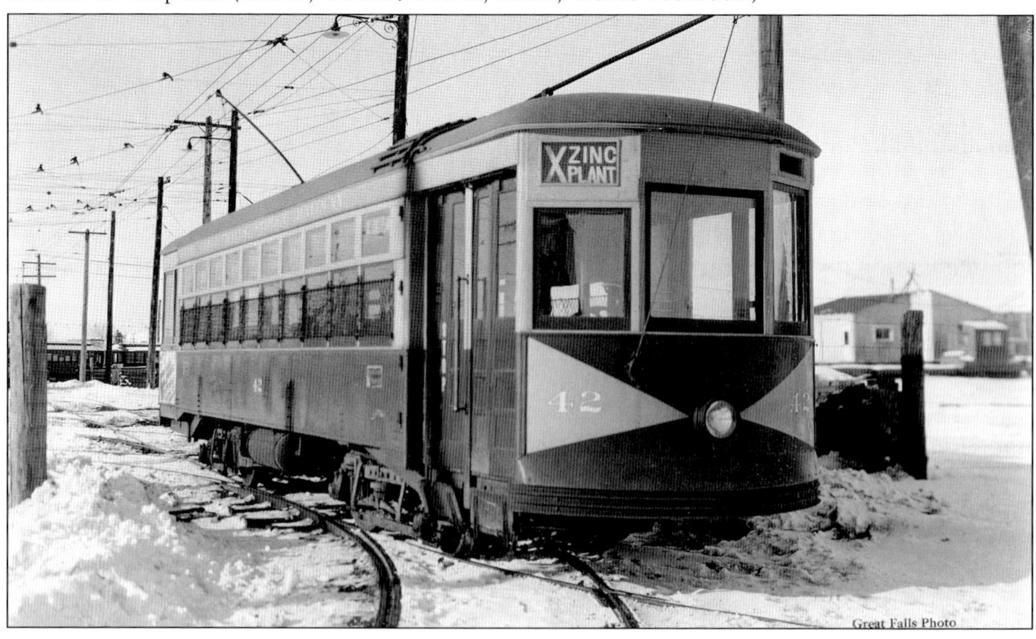

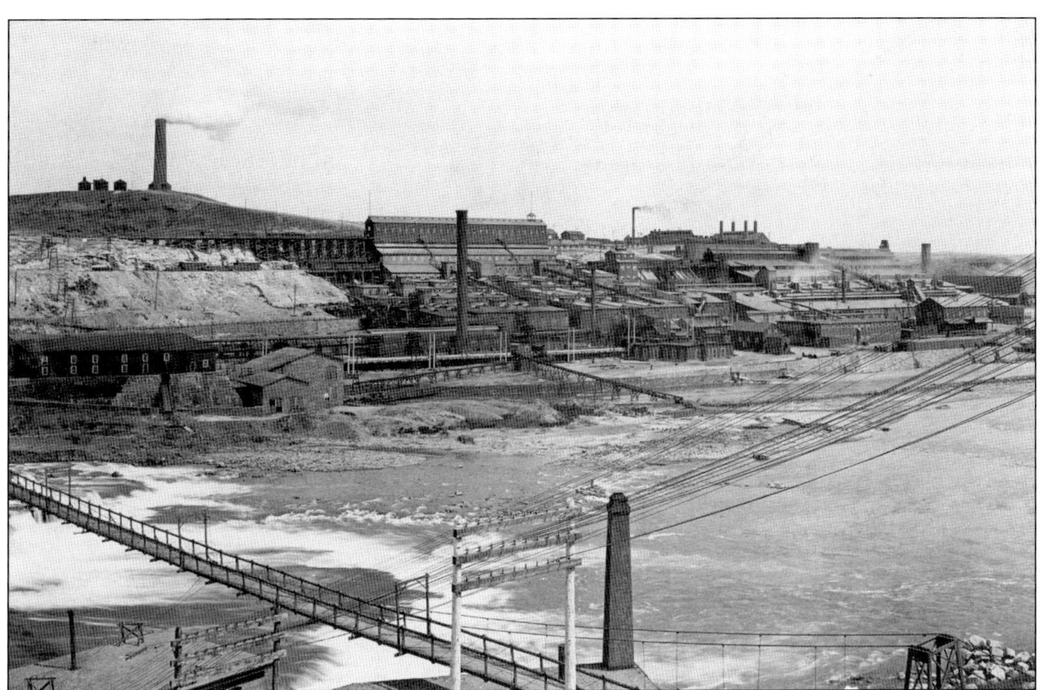

By 1901, Amalgamated Copper owned about 90 percent of the stock in the Boston and Montana Company and sought to consolidate it. This led to a four-year-long legal battle that Amalgamated won, turning Boston and Montana into a subsidiary. When all the dust settled, Amalgamated set aside $1.5 million to enlarge the smelter a little at a time. By 1905, the Great Falls Smelter was the third-largest producer of copper in the world. Both images here show the enlarged smelter. (Above, 2003.080.0028; below, 1982.017.0128.)

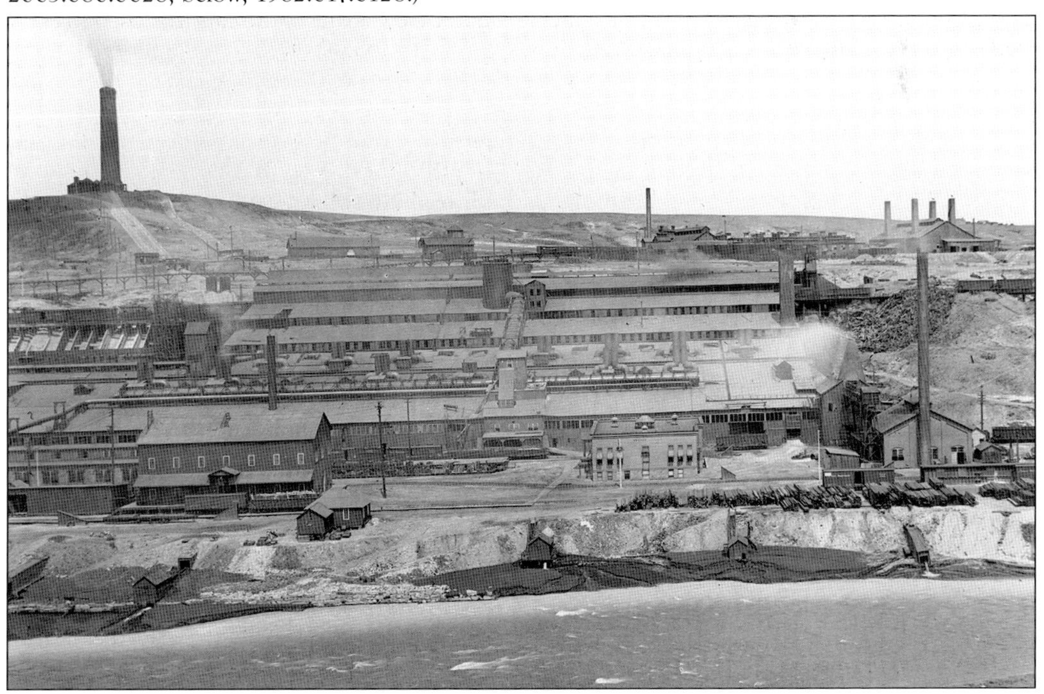

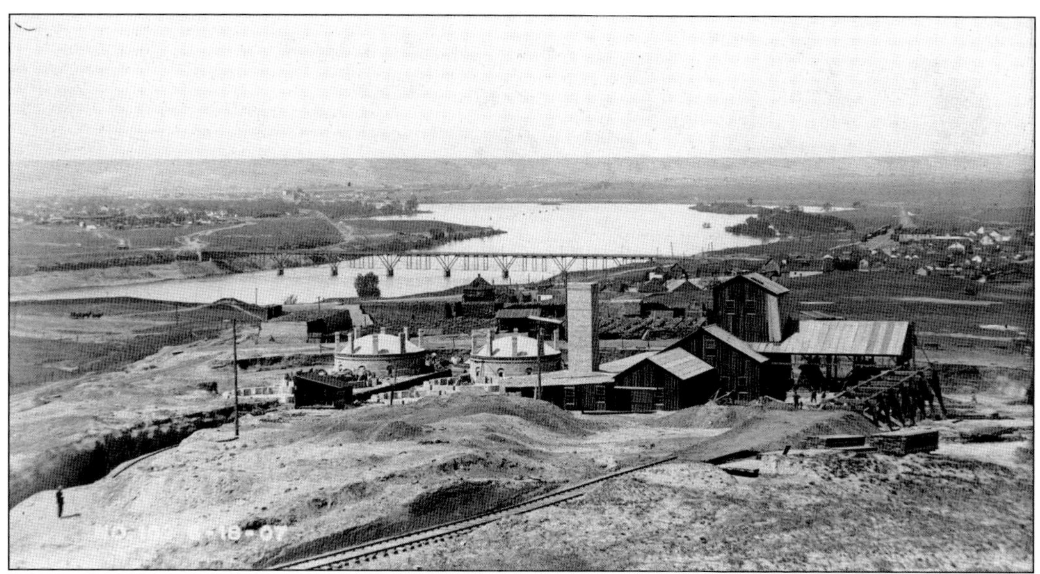

In 1906, construction began on an even larger expansion of the smelter, which also increased the need for coal from the nearby mines in Belt and Sand Coulee and required more stone from the quarry at Albright. As there was a suitable deposit of shale at the site, Combs and King built a brick plant to produce all the brick for the smelter's needs. It is seen here from above on June 18, 1907. It was then sold to the Anaconda Copper Mining Company in 1911 for over $45,000. It served the smelter and the surrounding community until 1932. It could produce 120 tons of various sizes of perforated radial bricks a day. Its capacity was put to the test for one of the most important improvements, the new smokestack. The stack was built by the Alphonse Custodis Construction Company of New York. The work of clearing the land and laying the foundation occurred from December 1906 to July 1907. Below is an image showing the different kinds of brick that were used for building the stack. (Above, from 2015.080.0001; below, from 2015.089.1.)

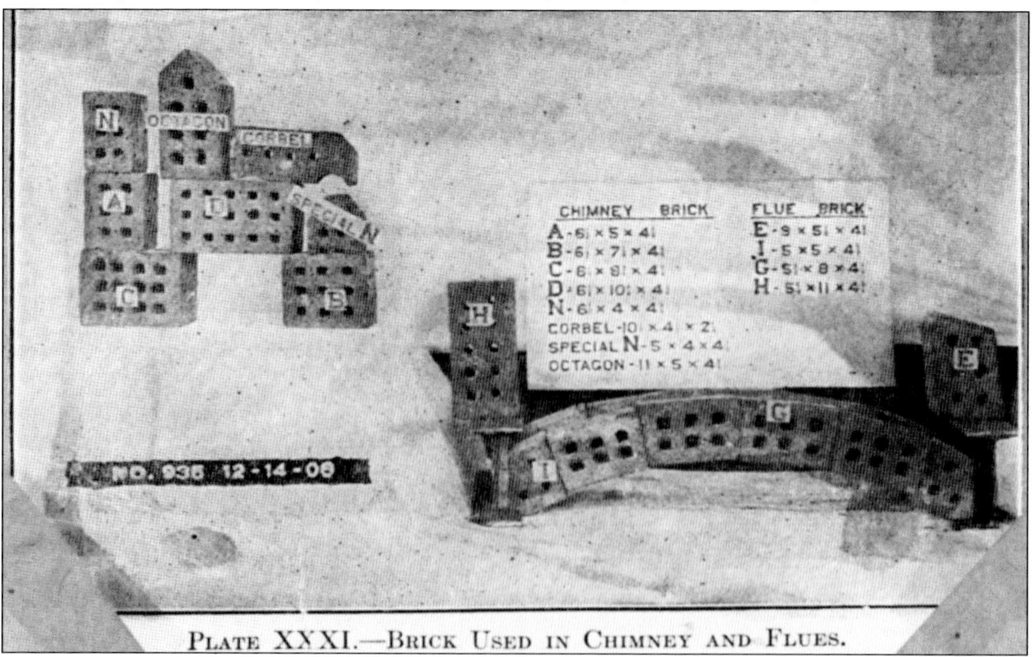

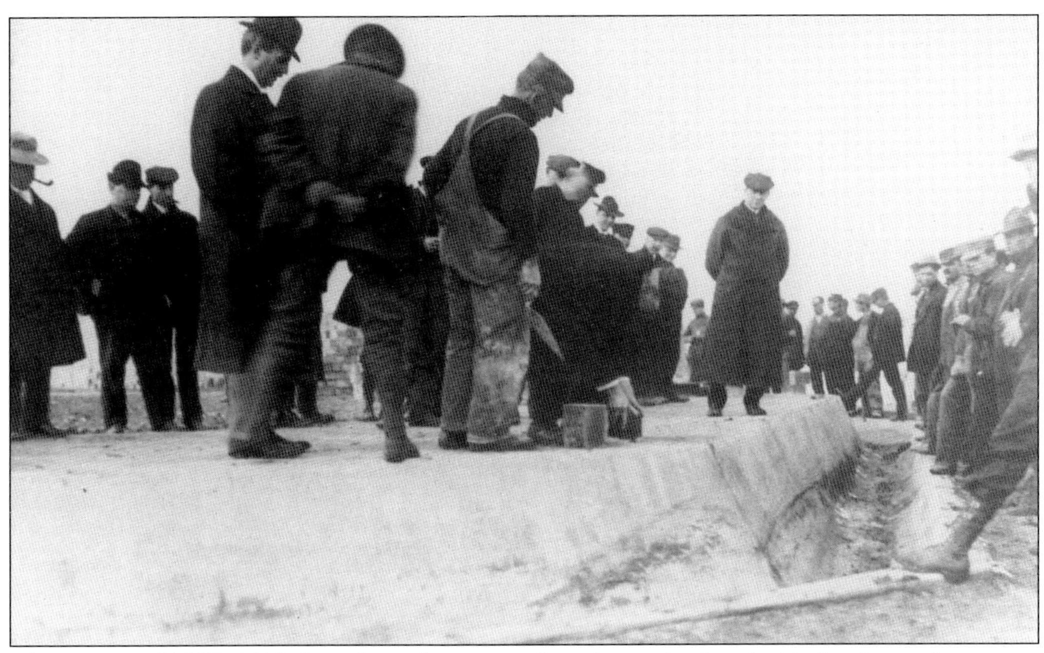

The winter of 1907–1908 slowed production and work, so the first brick for the stack was not laid until April 7, 1908. There was much pomp and circumstance for the event. The above image shows Benjamin B. Thayer, a representative for the president of Amalgamated Copper, placing the first brick. Below, John D. Ryan, managing director of Amalgamated, places the second. The third was laid by the superintendent of the Boston and Montana properties, C.E. Goodale, and the fourth was placed by A.E. Wheeler, superintendent of the smelter. (Above, 1988.112.0001; below, 1996.098.0010.)

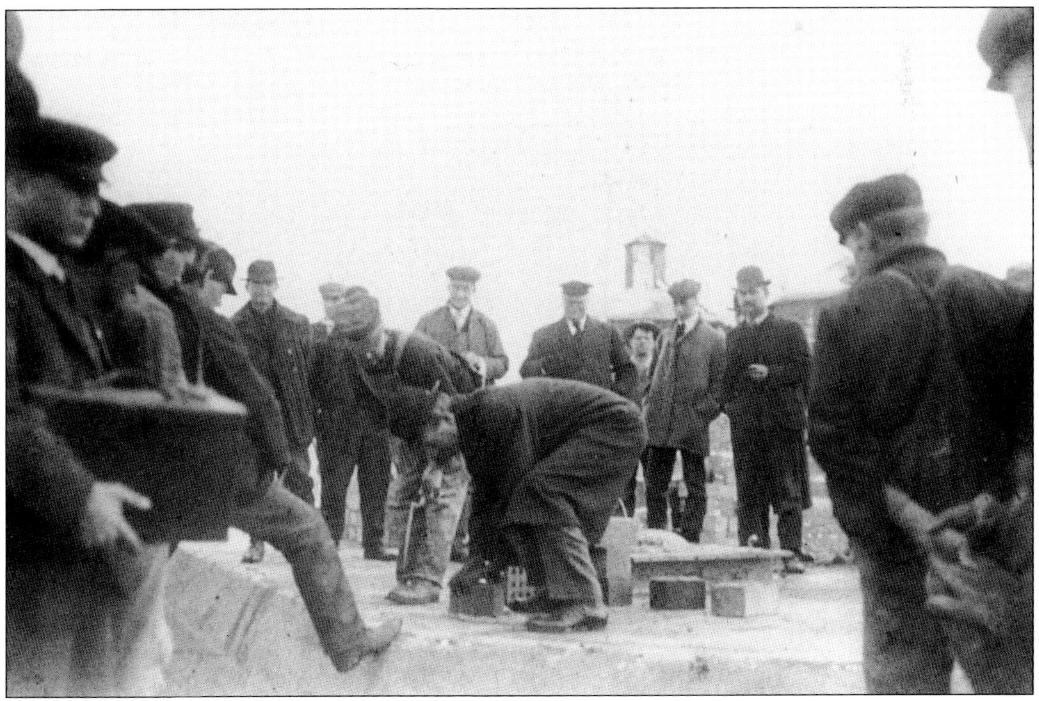

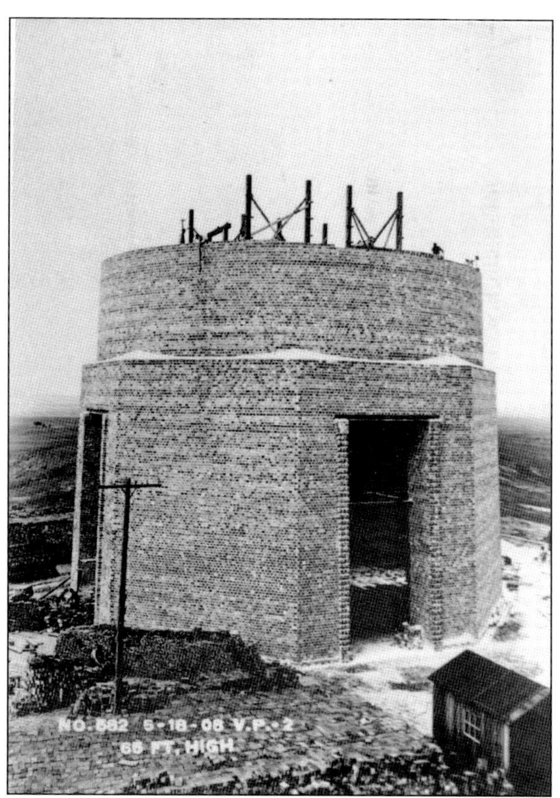

The stack construction employed 100 laborers, including 40 expert high chimney men. The workers also constructed a new flue measuring 1,800 feet to connect the furnaces to the stack. The image at left shows the first 66 feet of the new stack on May 18, 1908. Below, Conrad Worms, president of Alphonse Custodis Co., entertains guests on the top of the stack. Poker was played, Dutch lunch was served, and local American Brewing Co. beer was drunk. From left to right are Mike Horan, Post Office Newsstand; Conrad Worms; Fred Fligman, Paris Drygoods store; Sid Willis, The Mint Bar; Walter Clark, Merchants Association; Percy Raban, police magistrate; and Theodore Gibson. The men on the wall in the back are Charles Frenz and D.J. Sprague who worked on the stack and helped with the party. (Left, 2007.017.0275; below, CCHS 074.0006.)

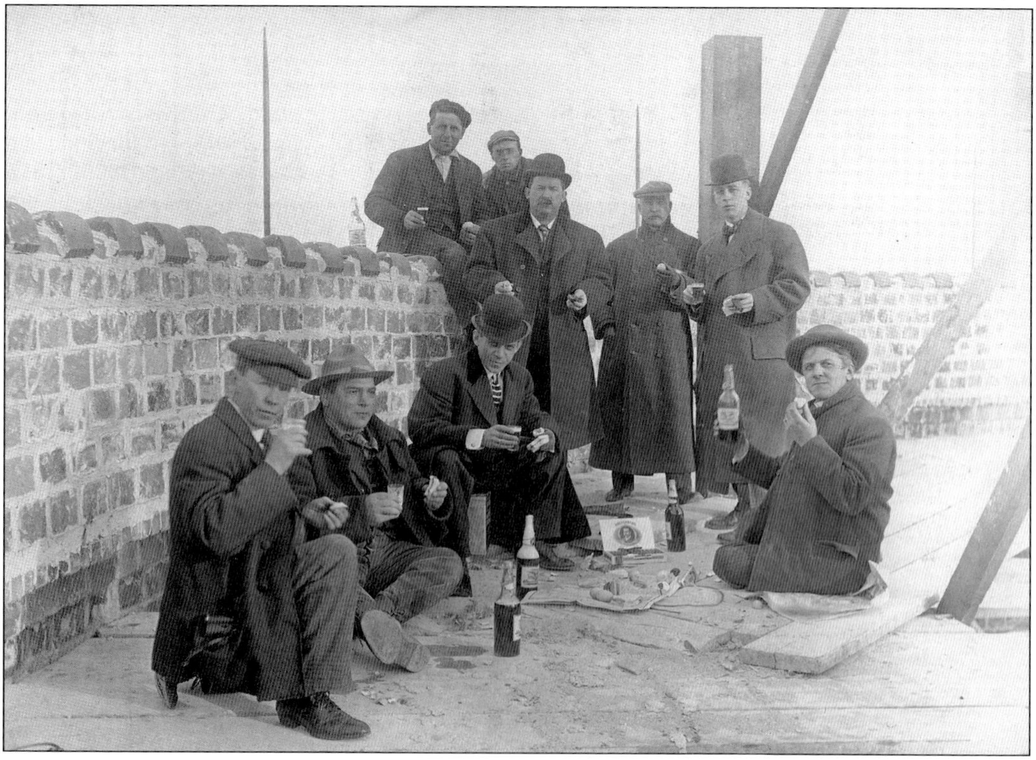

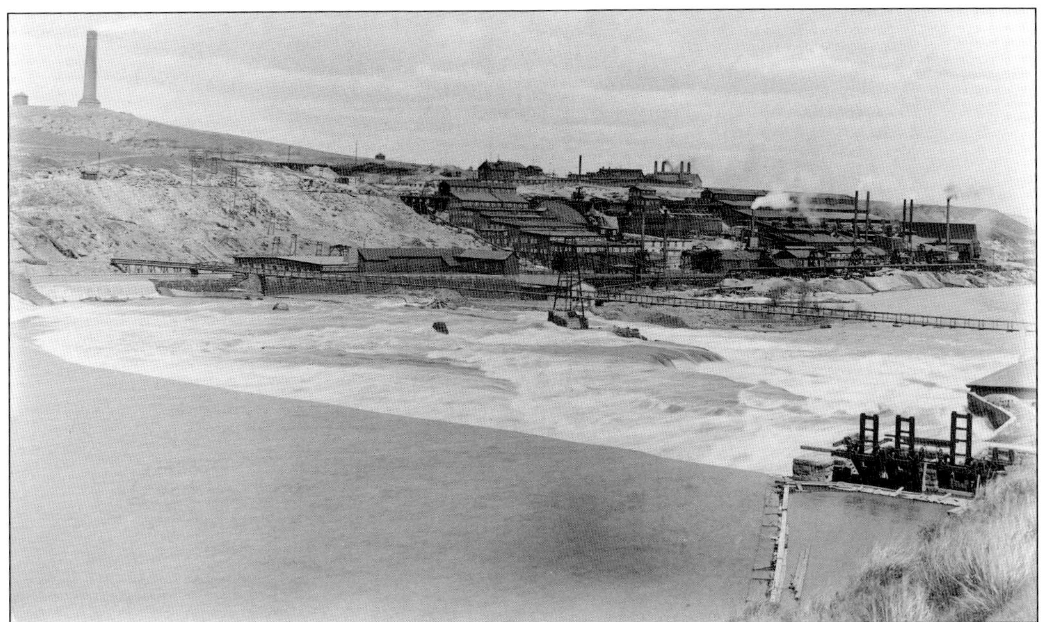

The year 1908 was a time of great prosperity for both the Boston and Montana Smelter and Great Falls. But it had its ups and downs, beginning with the largest flood Great Falls had ever seen. Starting June 5, it began to rain. It continued to rain, nearly three inches in 24 hours. The Missouri River swelled, and over the next five days, it became a massive flood, taking out homes, farms, bridges, and anything within its path. The substantial floodwaters are visible in these images. (Above, 2007.017.0278; below, 2007.017.0297.)

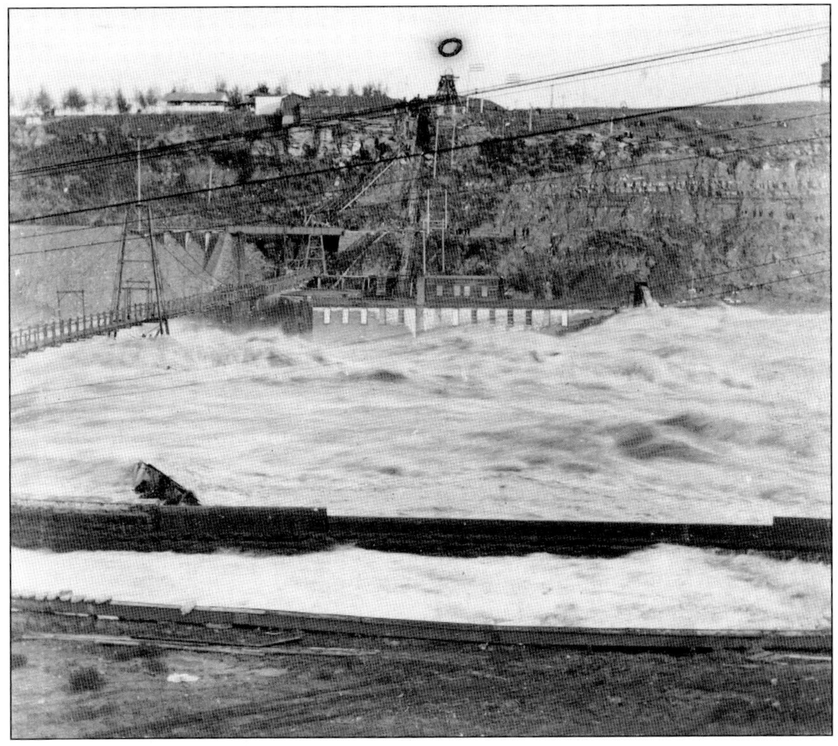

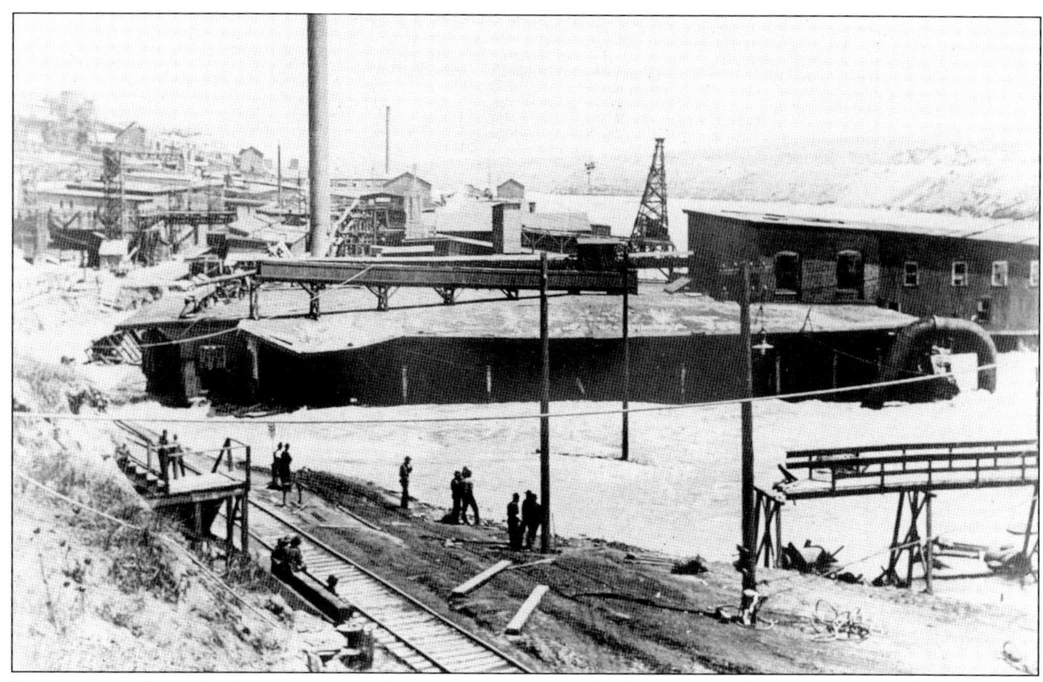

By noon on June 6, the suspension bridge connecting the Boston and Montana Smelter to Great Falls had to be closed off. Water rushed over the Black Eagle Dam, smashing floating buildings and other debris against it. At the peak of the flood, the water was 11 feet above the dam. The image above shows the floodwaters around the lower buildings. A few minutes after 2:30 p.m. on June 6, the north wall of the headrace that ran between the gatehouse and the powerhouse collapsed, flooding the powerhouse. Below is the gatehouse flooded through. All the workers escaped to the roof and watched as the building was washed away. Within three or four minutes, the whole plant shut down. (Above, 2019.015.0058; below, 1989.137.0161.)

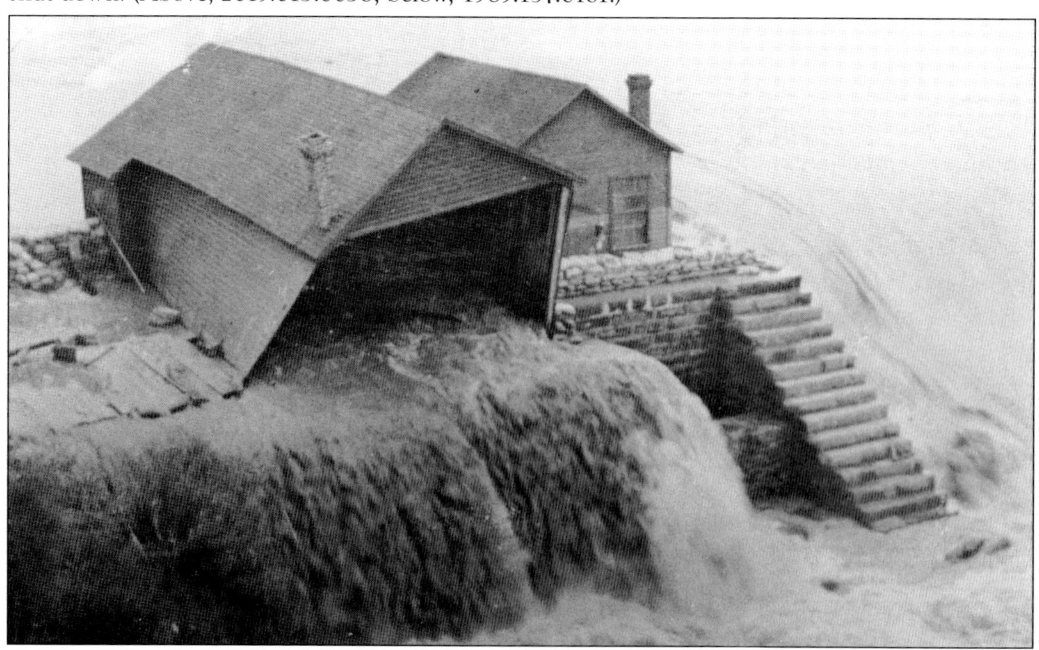

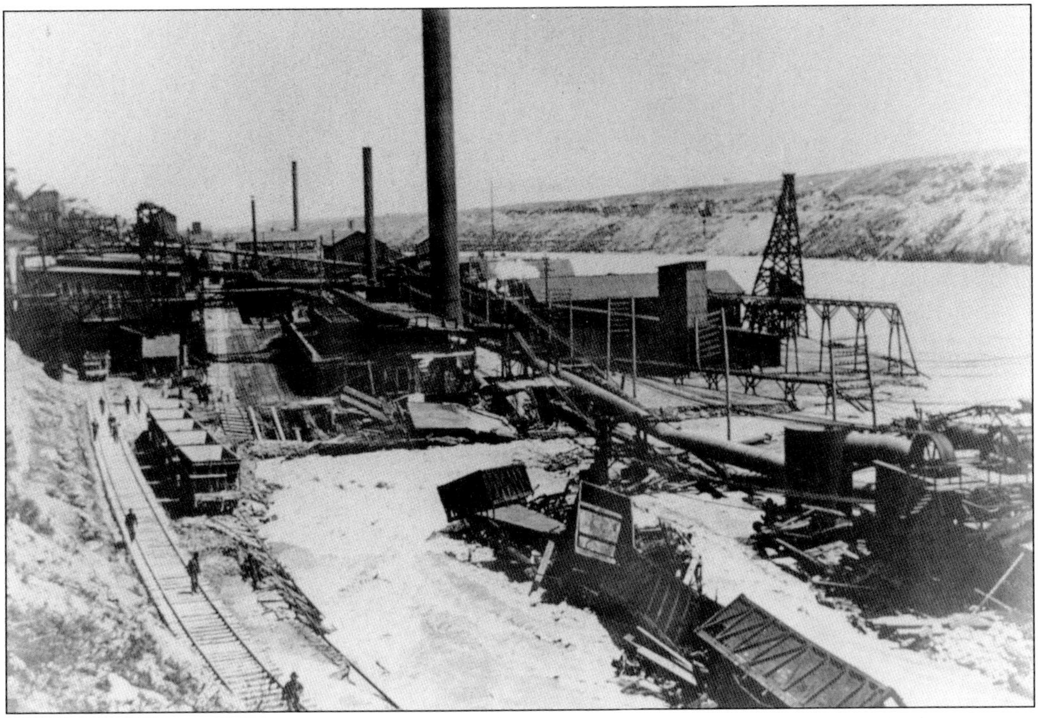

Despite constructing timber and rock barricades to prevent waters from reaching the lower parts of the smelter, between 50 and 75 feet of the north wall of the headrace was wiped out completely on June 6. Ore cars were dumped into the river in an attempt to stem the encroaching waters, but tragically they only served to angle the current toward the smelter buildings. The Missouri River ripped away the middle support of the suspension bridge and shredded the timber of the walkway, leaving only the cables. Note the ore cars in the photograph above and the remains of the walkway in the photograph below. (Above, 2019.015.0059; below,1982.056.0004.)

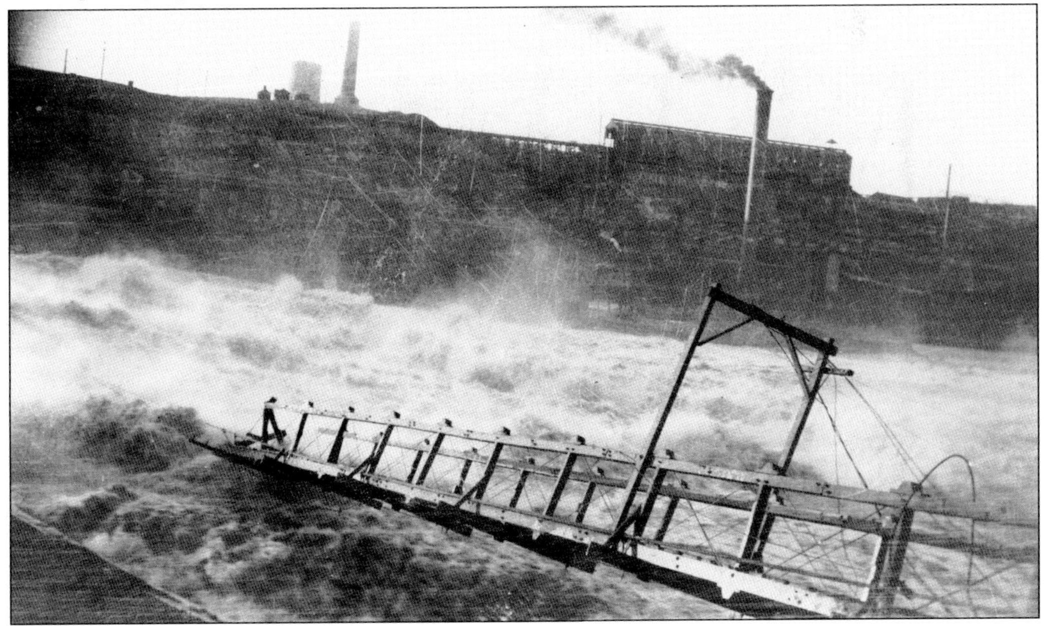

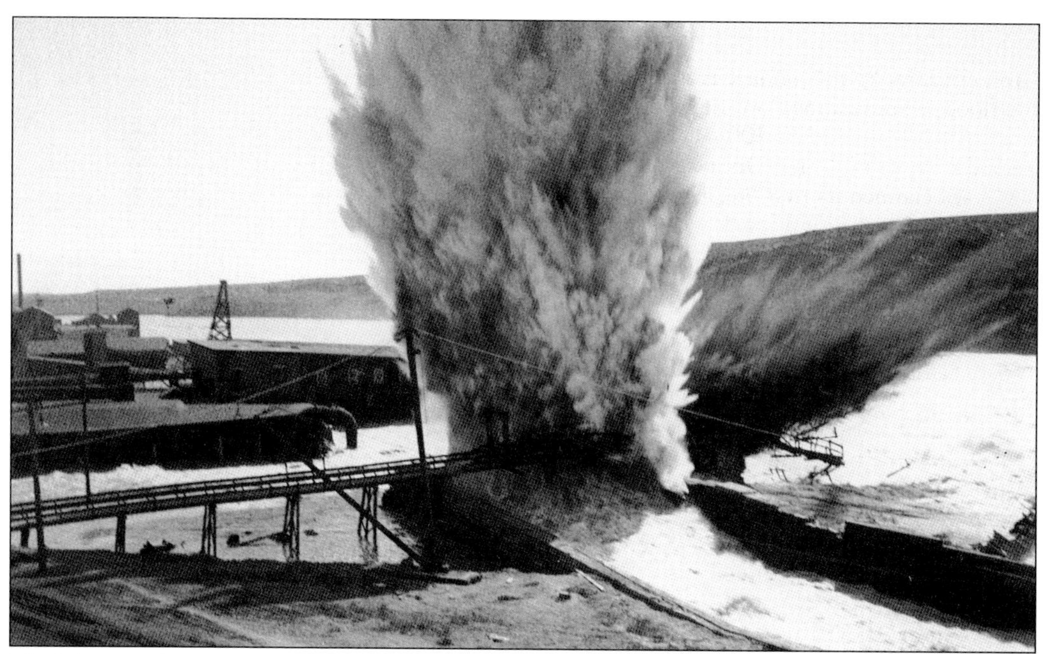

The Missouri River continued to eat more channels into the smelter grounds, washing away one of the pump houses, several coal cars, a boiler house, and parts of a trestle. A successful attempt to curb the channel of water was made on the ninth, when two sections of the south retaining wall of the headrace were blown out, as seen above, directing the water back into the main river flow. Not only were several of the buildings damaged, the railroad tracks and bridges along the Missouri, Sun, and Smith Rivers were all damaged. Telephone lines and communications were broken, isolating many from help or news. For those five days, there was a continuous roar that could be heard throughout the area. A closer view of the floodwaters is shown below. When the water finally receded, the damage was manageable. The water was finally shut out after two weeks of repairs to the walls and headrace gate. (Above, 1982.129.0002; below, 2007.017.0294.)

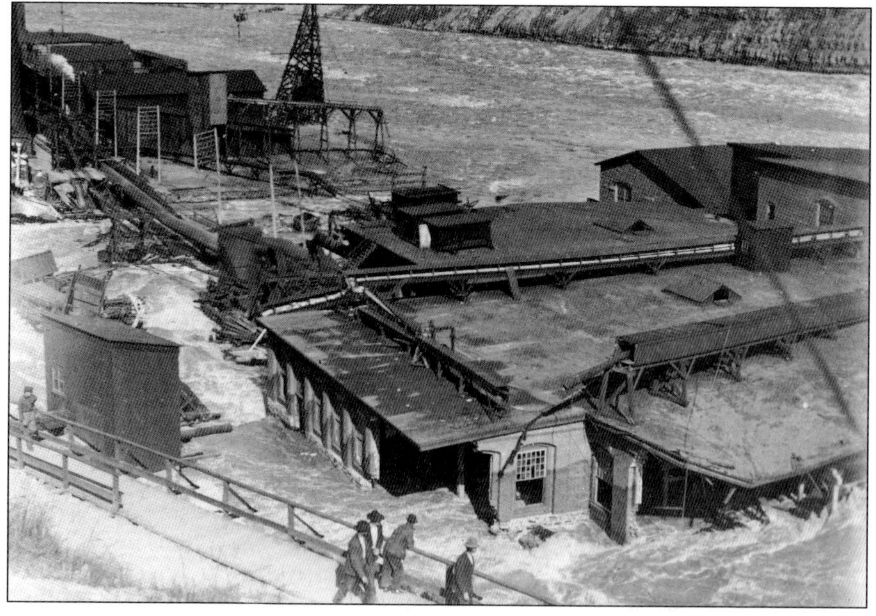

After cleaning up the damage from the flood, work resumed on the new stack. At right is the first 400 feet of the stack during its building. On July 9, the new stack claimed its first life, a young bricklayer named Ralph Jones, who at about 10:15 a.m. took a false step and fell from his scaffold 180 feet to the ground. He tried to grab some of the scaffolding on his way down, but it just put him into a somersault. He was terribly broken and bruised but conscious upon hitting the ground. He was taken to Columbus Hospital, where he was declared dead on arrival. Another life was claimed in May 1909. Peter C. Yeager was helping to remove the scaffolding from the inside of the stack as part of the lining process. While he was knocking a timber loose, the one he was standing on either tipped or gave way. He fell 220 feet, his body being brutalized by scaffolding on the way down. He died before he hit the ground. The finished stack is seen below. (Right, 1992.066.0019; below, 1997.081.0001.)

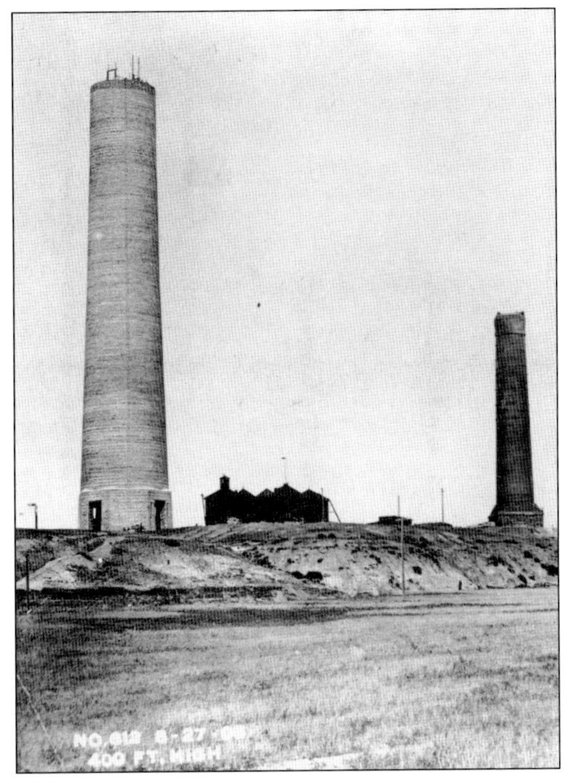

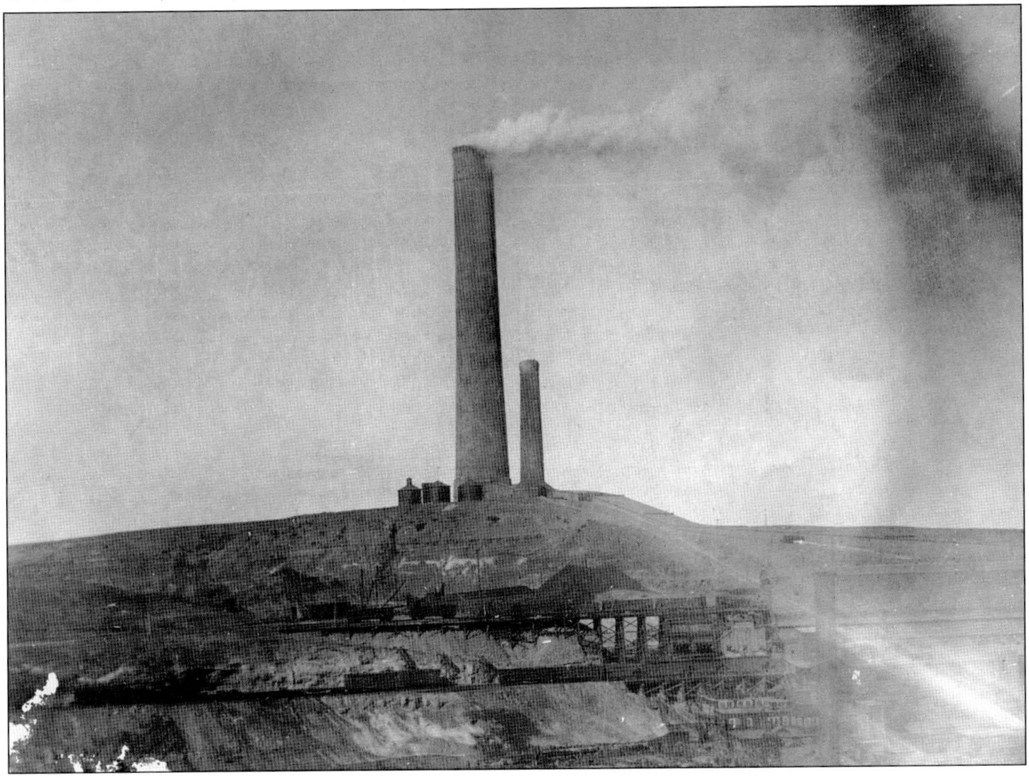

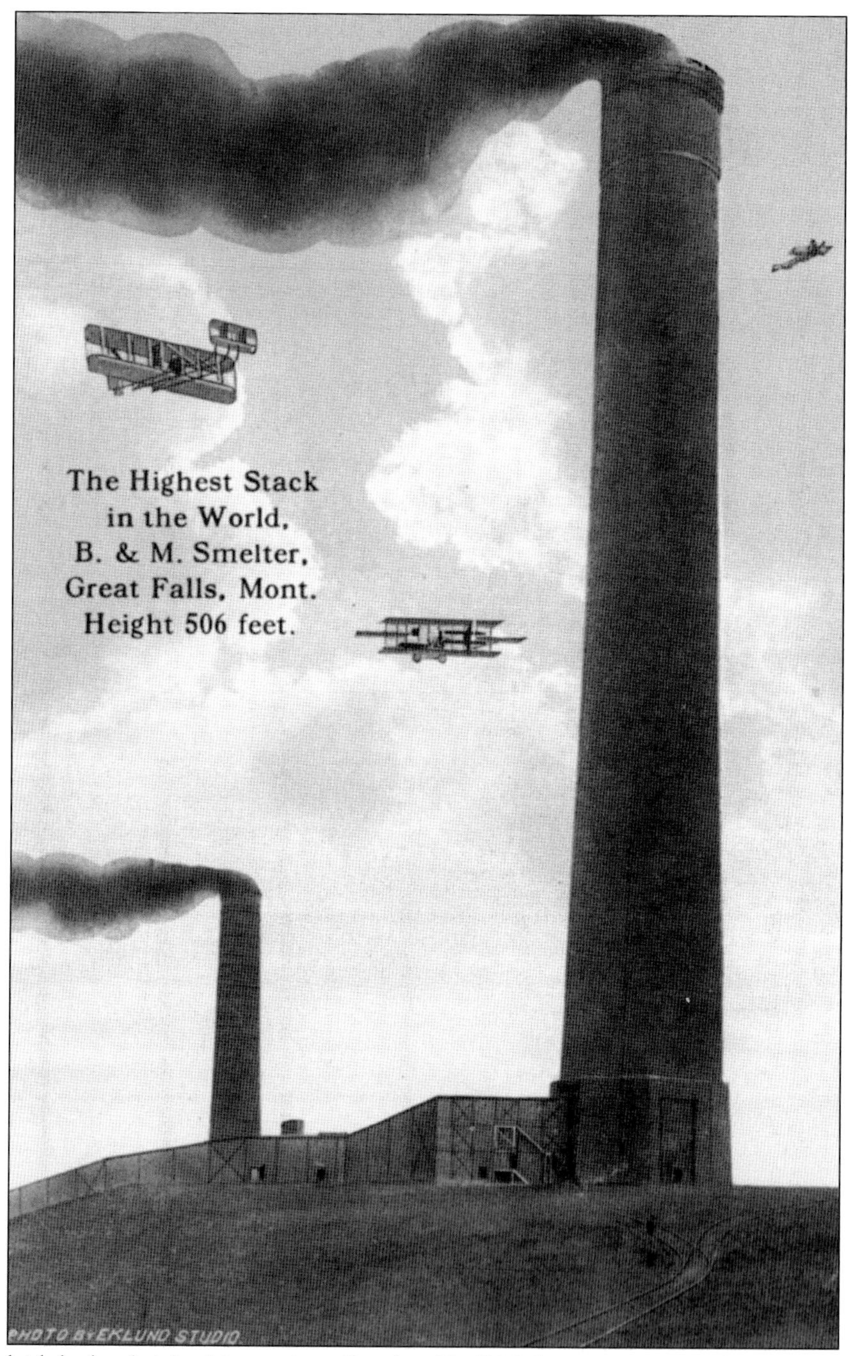

Workers laid the last brick of the stack on October 23, 1908, but it was not until June 12, 1909, after all the blood, sweat, and sacrifices, that the smokestack issued its first smoke. At its completion at 506 feet, it was the tallest smokestack in the world. The "Big Stack" quickly became an icon not only of the city of Great Falls but of Montana industry. In 1910, Amalgamated and several other companies merged, and the most powerful company the state had ever seen was born, Anaconda Copper Mining Company. The local smelter became known as the Great Falls Reduction Department. Above is a postcard featuring the new world's tallest smokestack at 506 feet tall. (2018.049.0017.)

Two

The Process for Copper

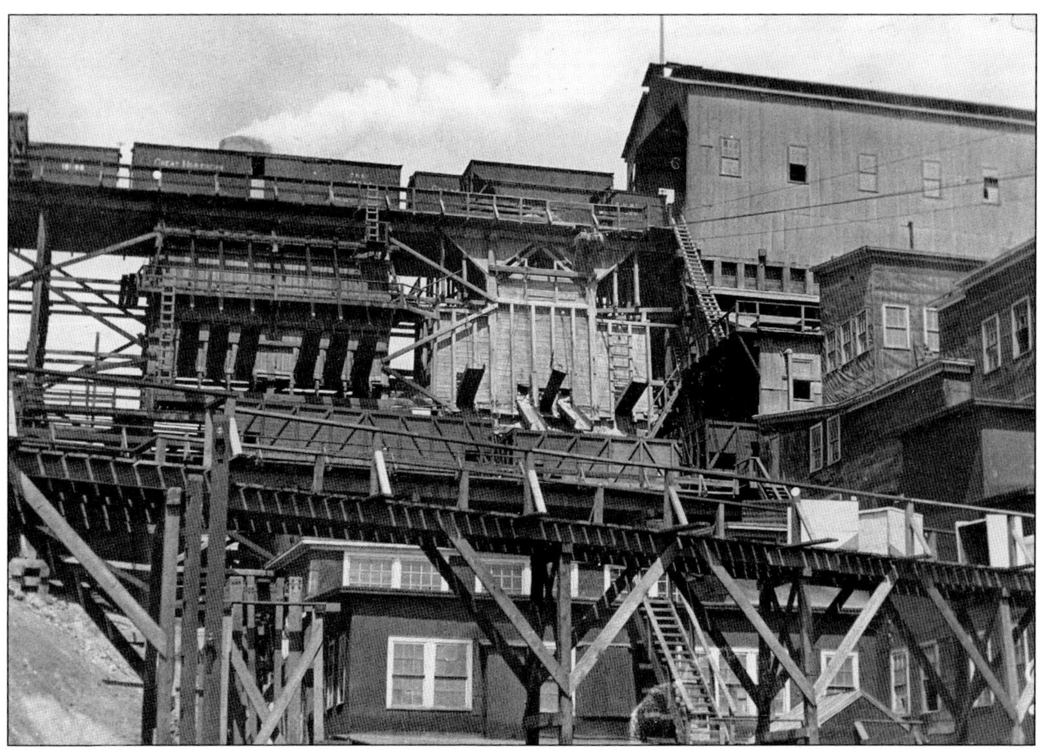

The first copper refinery plant was built in 1892. At that time, copper ore mined in Butte, Montana, would be sent up by train to the smelter, like in the photograph above. The ore would be concentrated, roasted to remove the sulfur, sent to the reverberatory furnaces, and then treated in the converters. From there, the ore was ready to melt into anodes and be sent to the Electrolytic Refinery Plant. (1993.111.0027K.)

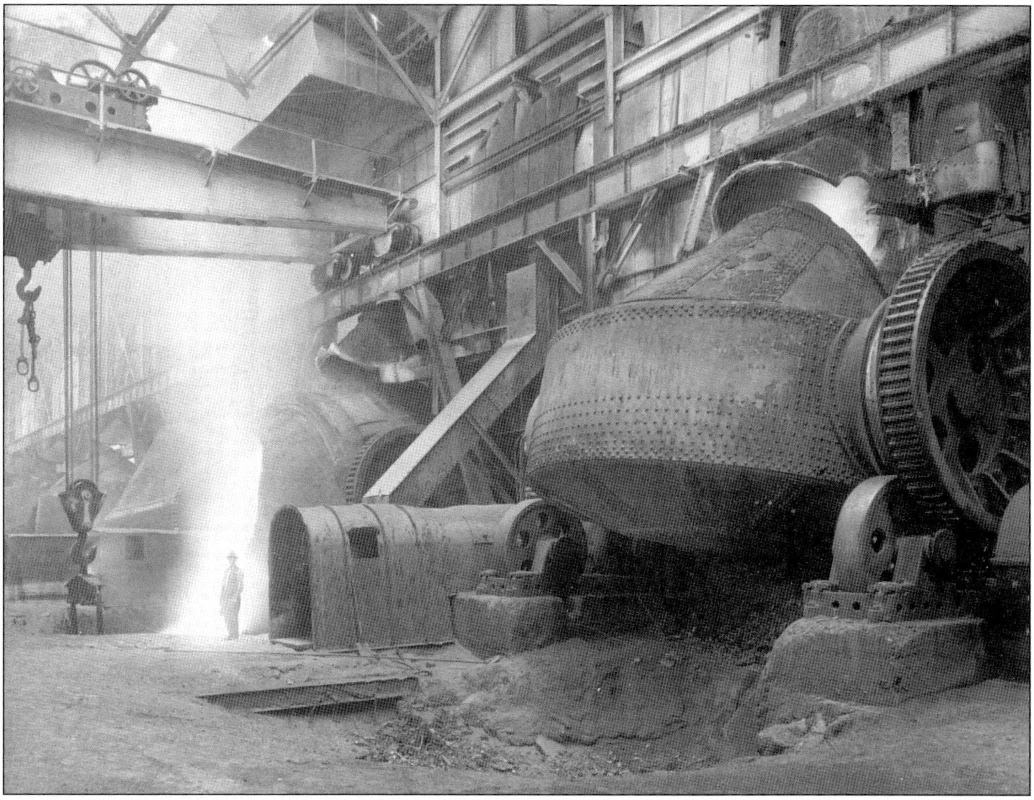

In 1916, a new Copper Refinery Plant was built at Great Falls, as seen below. However, in 1918, a new reduction works was built at Anaconda, Montana, the Washoe Smelter, and the entire processing of ore to anodes was moved there. Above are the copper concentrators that were at the smelter in Anaconda. At that point, copper came to the Great Falls plant as anodes ready for the Electrolytic Refinery Plant. The plant was no longer a smelter but a refinery. (Both, from 2015.089.1)

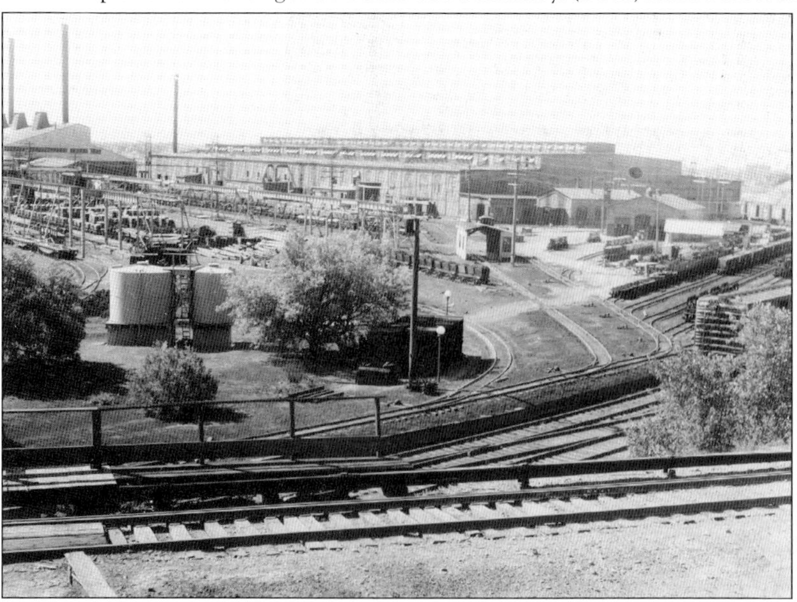

The anodes were 99.3 percent copper and weighed 630 pounds each. The other 0.7 percent was made up of silver, gold, arsenic, and tiny amounts of miscellaneous elements. They arrived in boxcars and were moved with chains one at a time into rack carts that held 25 each. At right, Harold Barlow (left) and Gene Marianetti (right) unload anodes, and below, Gene Marianetti places the anodes into the holding cars. Then they were moved by the internal company electric tram system into the Electrolytic Copper Refinery building. (Right, from 2015.089.1; below, 1982.017.0273.)

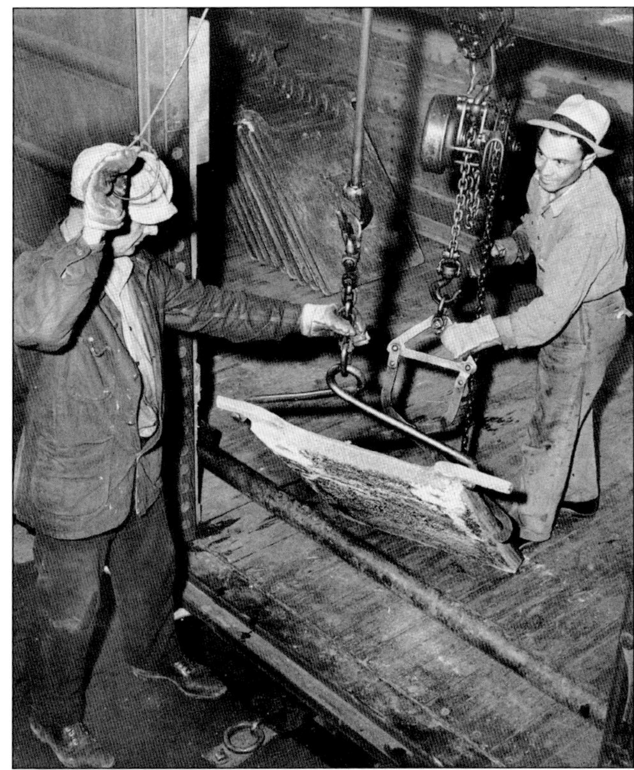

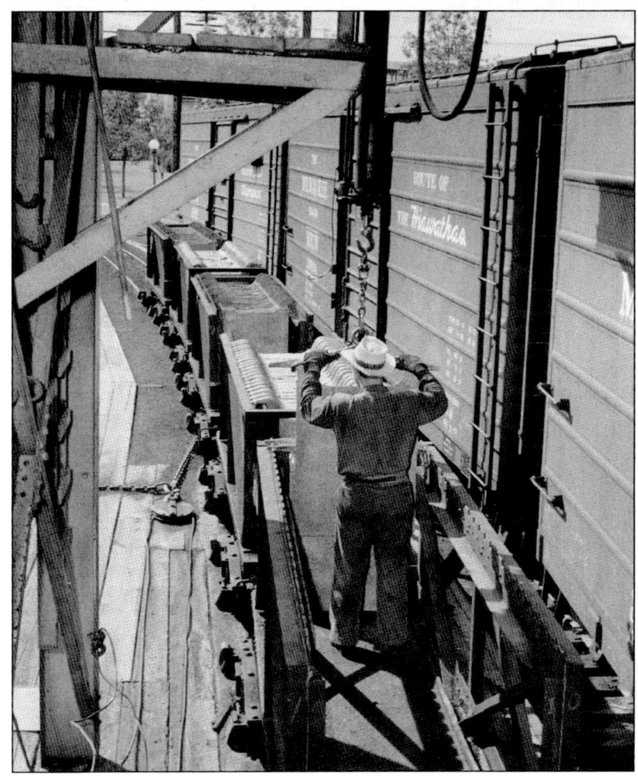

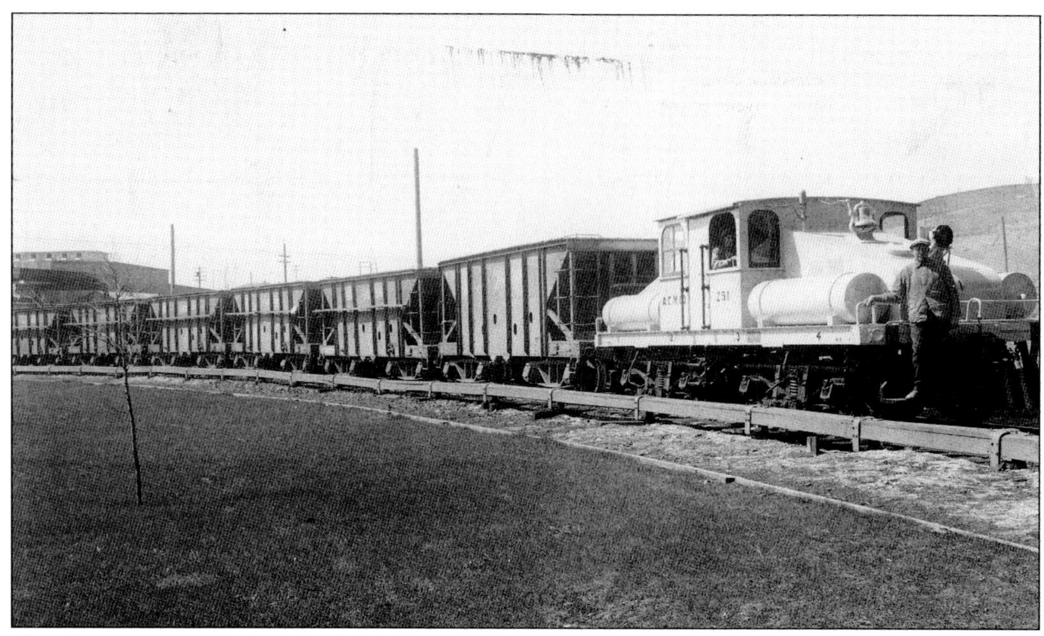

The electric tram system moved on average 2,000 tons of material daily. Above, one of the electric tram engines pulls cars. The grounds of the plant consisted of nine miles of standard-gauge track. The Electrolytic Copper Refinery building, seen below, was enlarged in 1926 and then was able to handle a capacity of 27 million pounds of copper per month. The main tank house expanded 152 feet more. Separate smaller buildings around the tank house are the office, shops, roundhouse, and slime plant. (Above, 1982.017.0289; below, from 2015.089.1.)

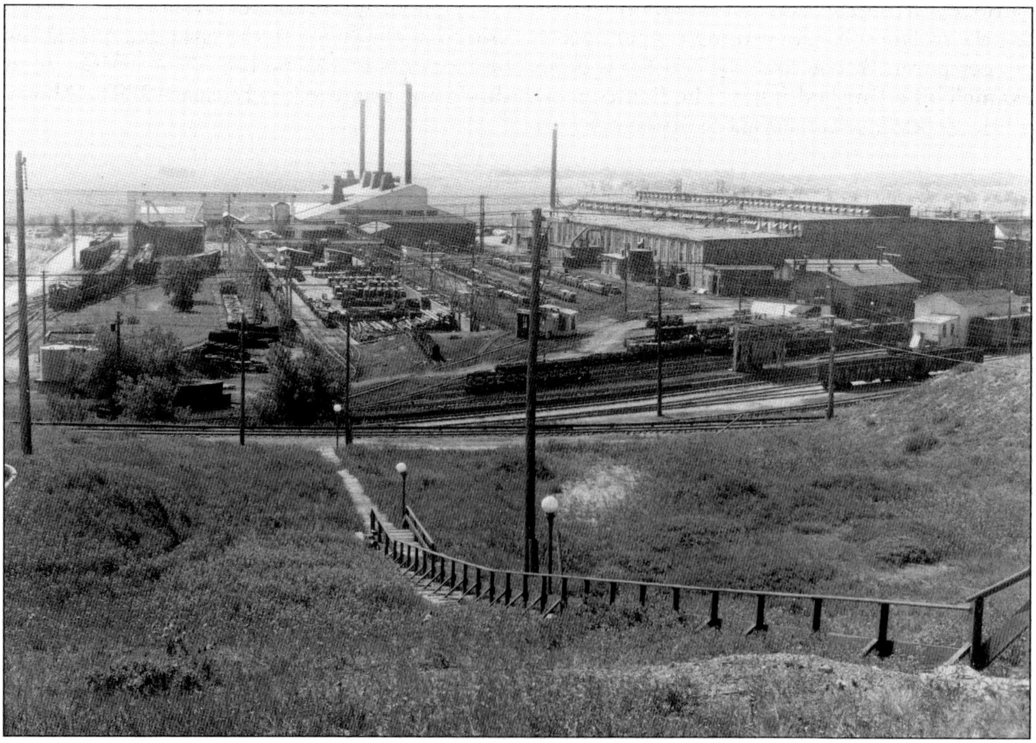

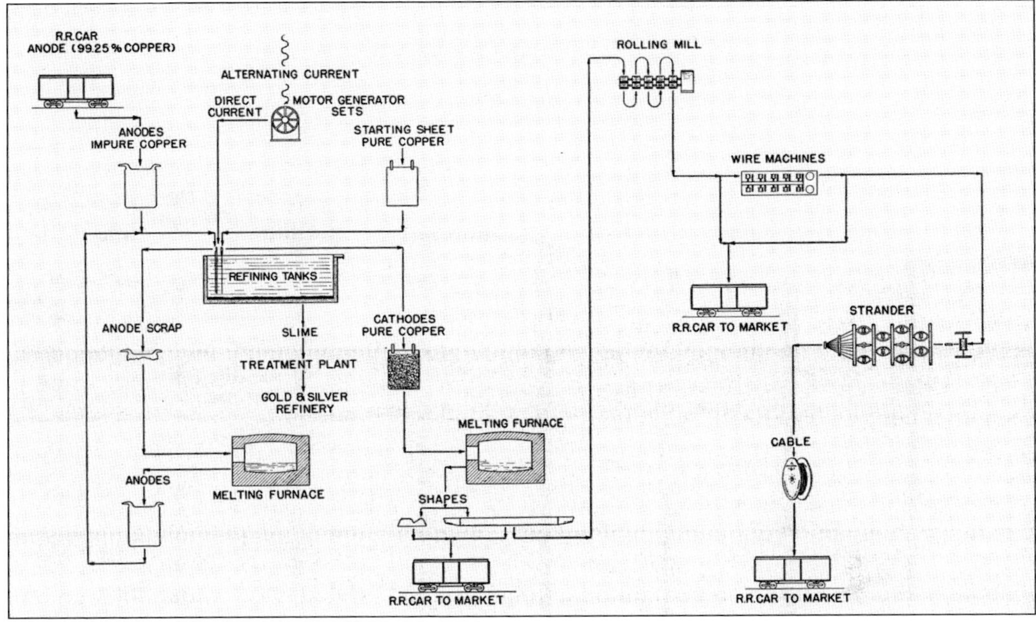

This diagram shows how the copper was processed once it arrived at the Great Falls Reduction Department. The copper went through the electrolytic refining process, where an anode and cathode starting sheet are put into refining tanks with electrolyte solution. Electricity runs through the electrolyte solution, causing copper ions to move from the anode to the cathode, making the copper purer. By the time the process is done, the copper is at 99.9 percent purity. Below is an example of a finished ingot. The first copper cathode was produced in February 1893. (Above, 1991.119.0027; below, 1982.017.0036.)

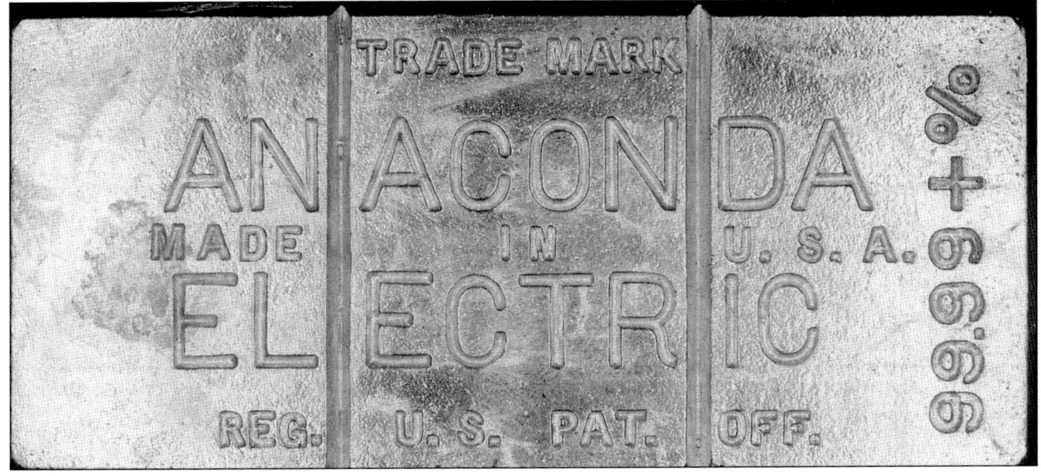

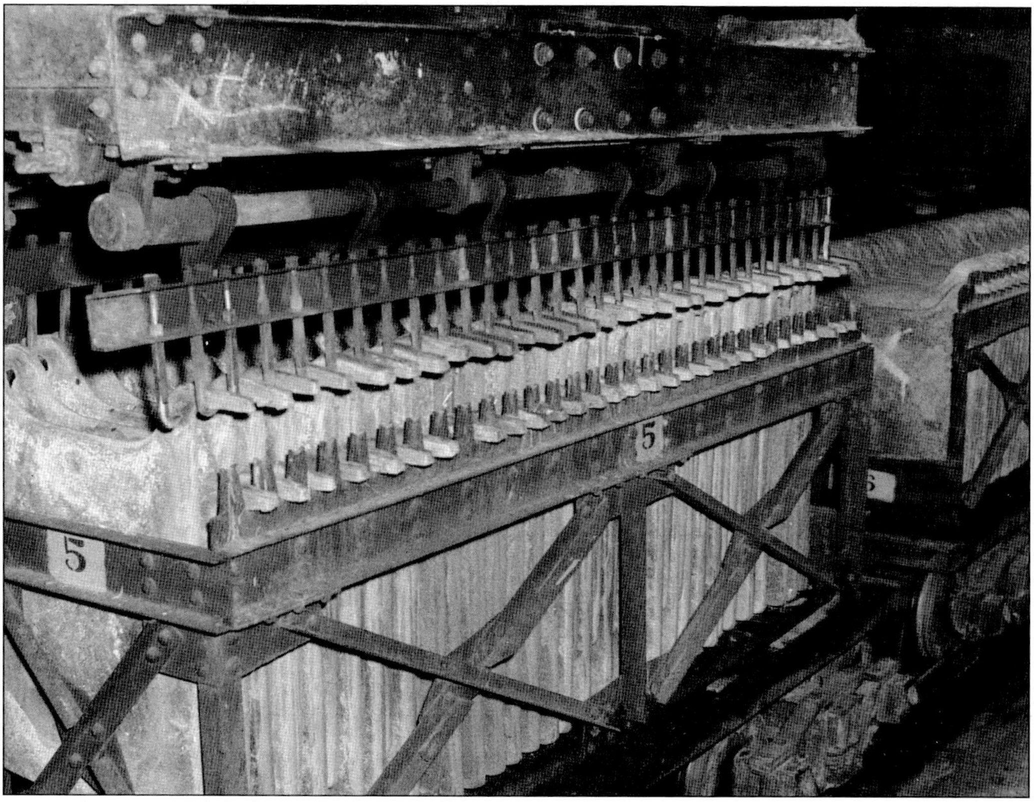

An air hoist moved the anodes from their rack carts to one of the refining tanks, as seen above and below. Each tank held 25 anodes and 26 cathodes. The anodes stayed for 24 days at a time, and the cathodes were drawn four times or every six days. The small crane hooks for lifting the anodes and cathodes were made from Everdur, a silicon brass that does not dissolve if it gets any of the electrolyte solution on it. (Both, from 2015.089.1.)

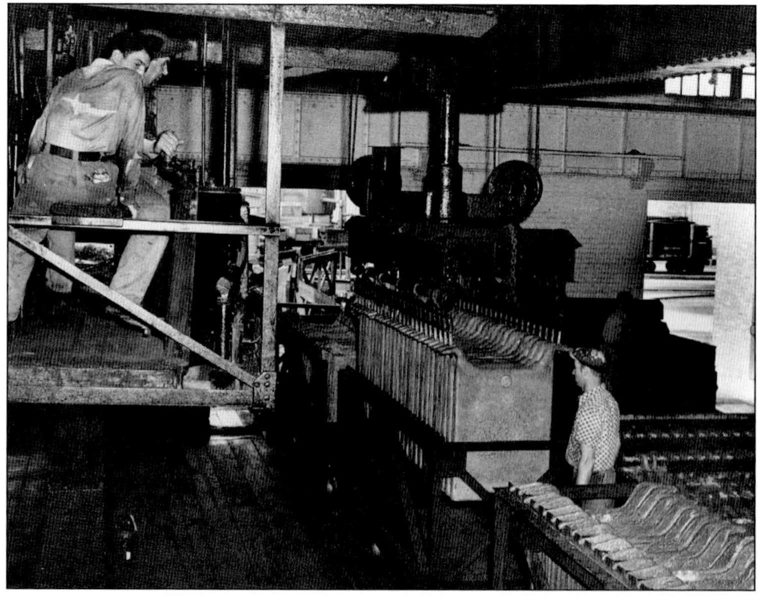

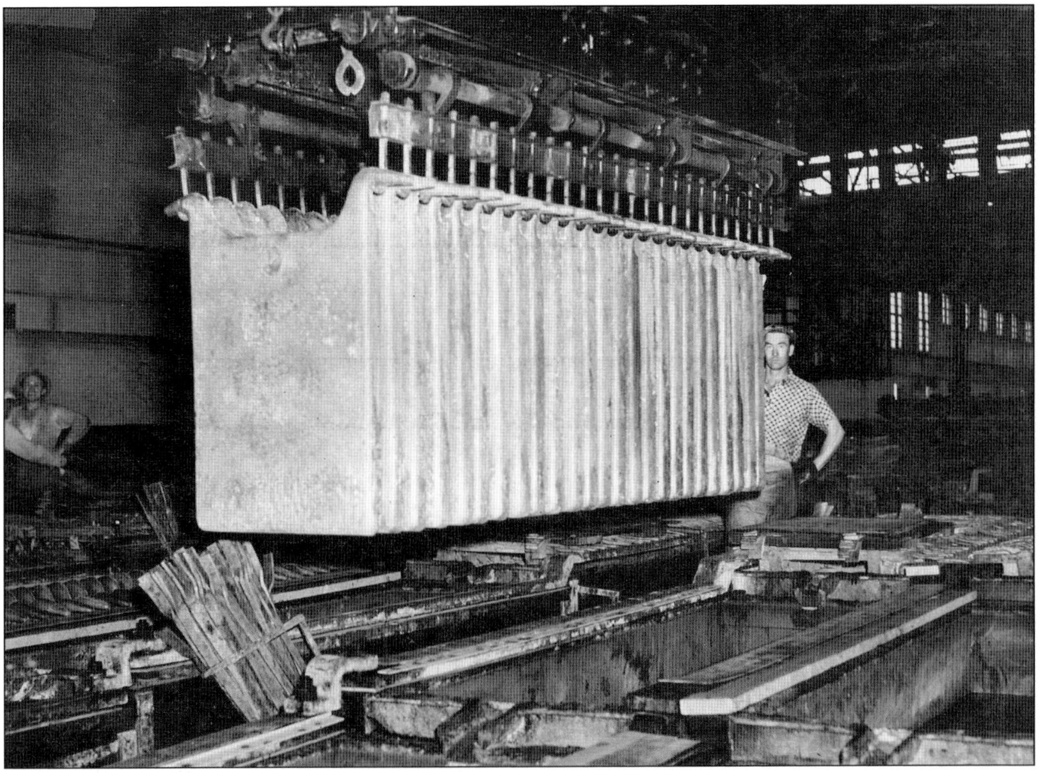

At the end of the 24-day period, what was left of the anodes, or anode scrap, was pulled and sent to the anode furnace for recasting. Intermittently, when the anodes were changed, the solution in the tanks was pumped out, and the slime went through a system of lead-lined launders to the collecting tank and then to the slime plant for drying and packing. The purity of the electrolyte was maintained by daily running some of the solution through the purification plant to prevent a buildup of impurities. Above, anodes are being loaded into the tanks, and below, the cathodes are being loaded in between the anodes. (Above, from 2015.089.1; below, 2008.099.0011.)

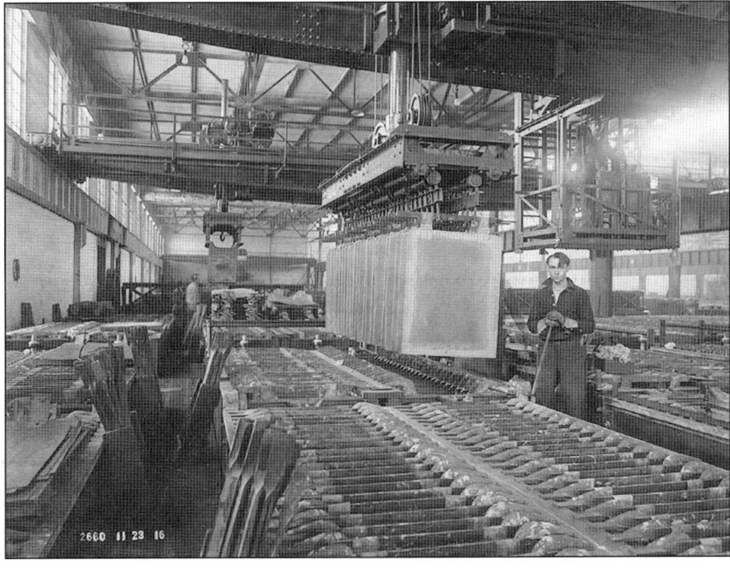

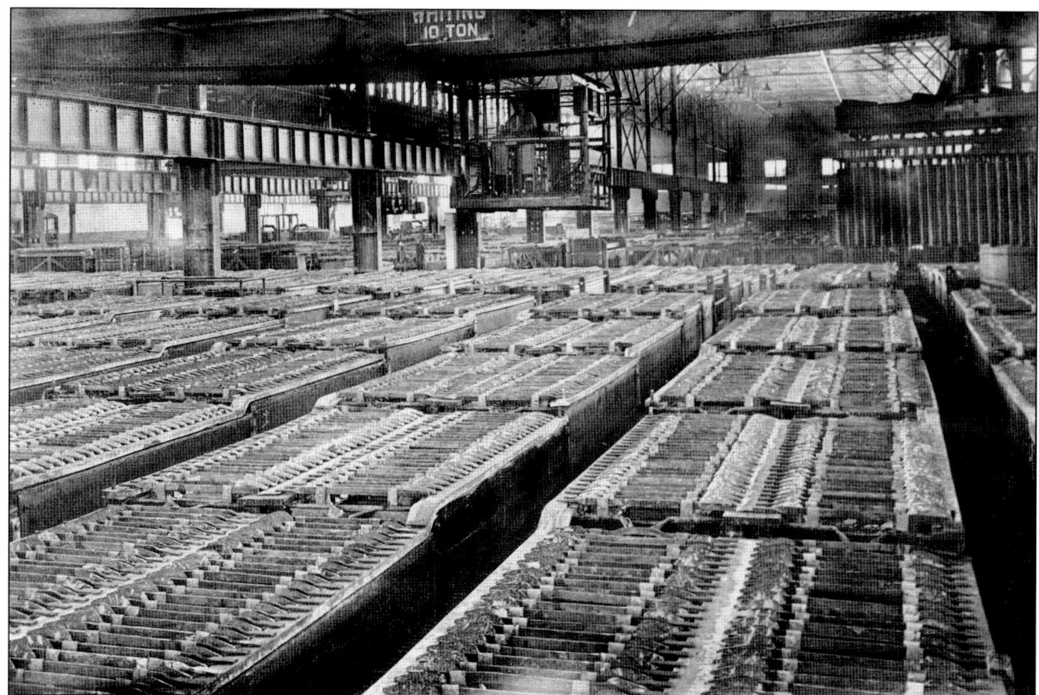

The tank house contained 1,530 refining tanks. The tanks themselves were 10 feet, 3 inches long; 2 feet, 10 inches wide; and 3 feet, 9 inches deep. They were made of Douglas fir wood and lined with eight-pound six percent antimonial sheet lead. The tank house was divided into four crane bays. Each section had its own separate electrolyte solution and circulated at a rate of five gallons per minute. Above is the inside of the Electrolytic Copper Refining Tank Room, and below is a closer view of how the anodes and cathodes are interleaved. (Above, from 2015.089.1; below, 1982.017.0168.)

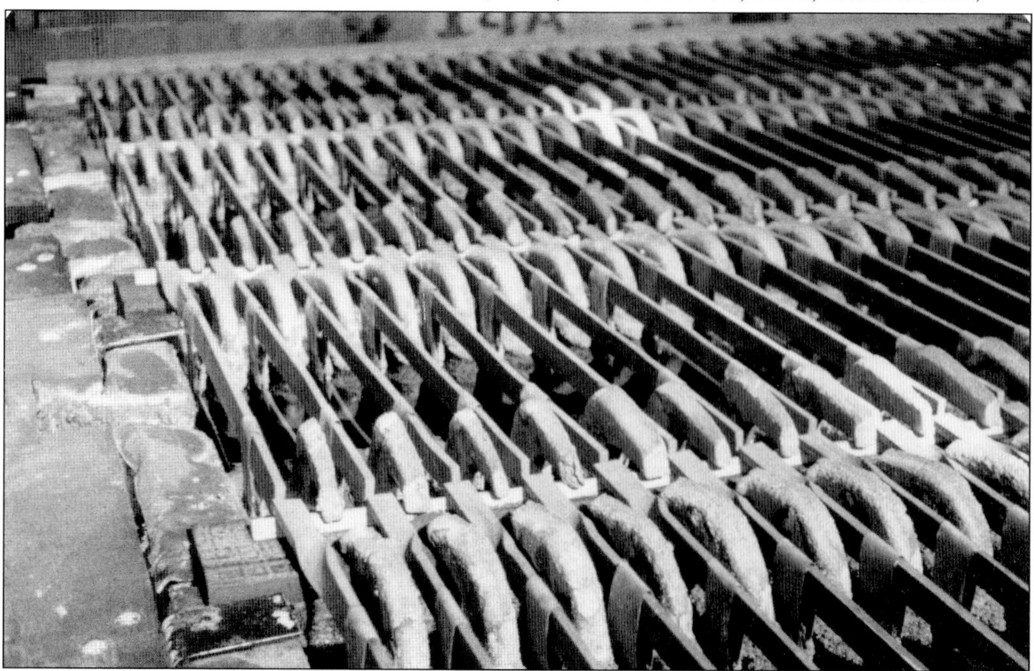

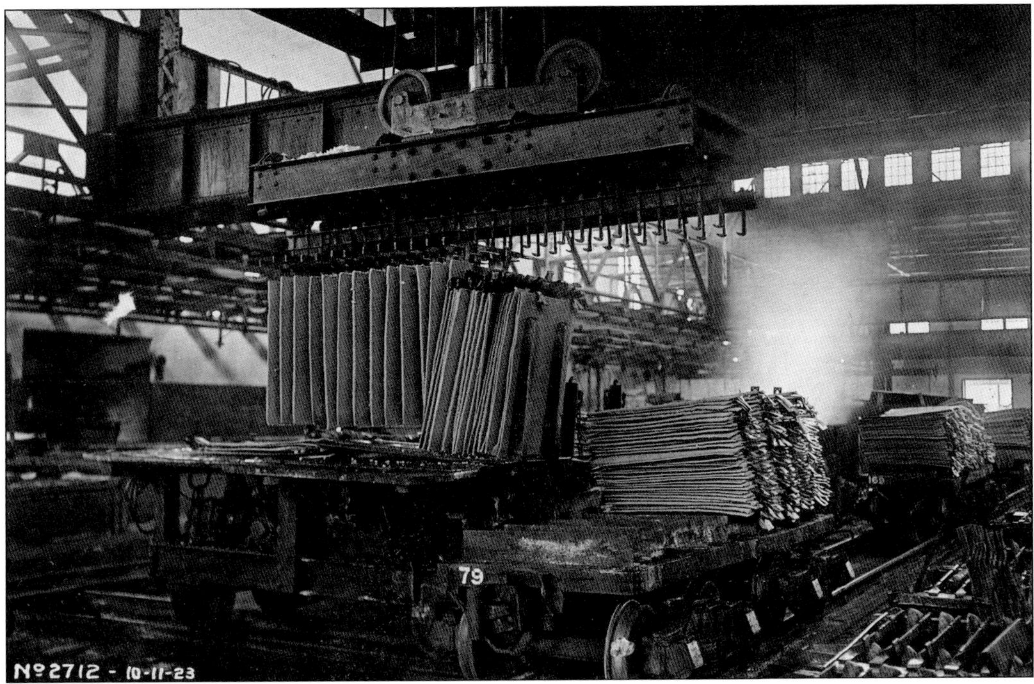

Cathodes were created in the tank room in their own separate tanks. A starting sheet was put on a copper blank. A coating of Texaco Nabob oil prevented the sheets from sticking to the blanks. Pine tar was applied to the edges to prevent the sheets on both sides from bonding together. They went into the electrolyte solution and then were stripped once enough copper has deposited on the sheets. From there, the sheets went to the sheet room and got their loops put on for their turn in the refining tanks, as seen above. In 1956, the refinery began a program of modernizing and increasing its capacity for producing cathodes. Below is a stack of finished cathodes. (Above, 1982.017.0316A; below, 1982.017.0166.)

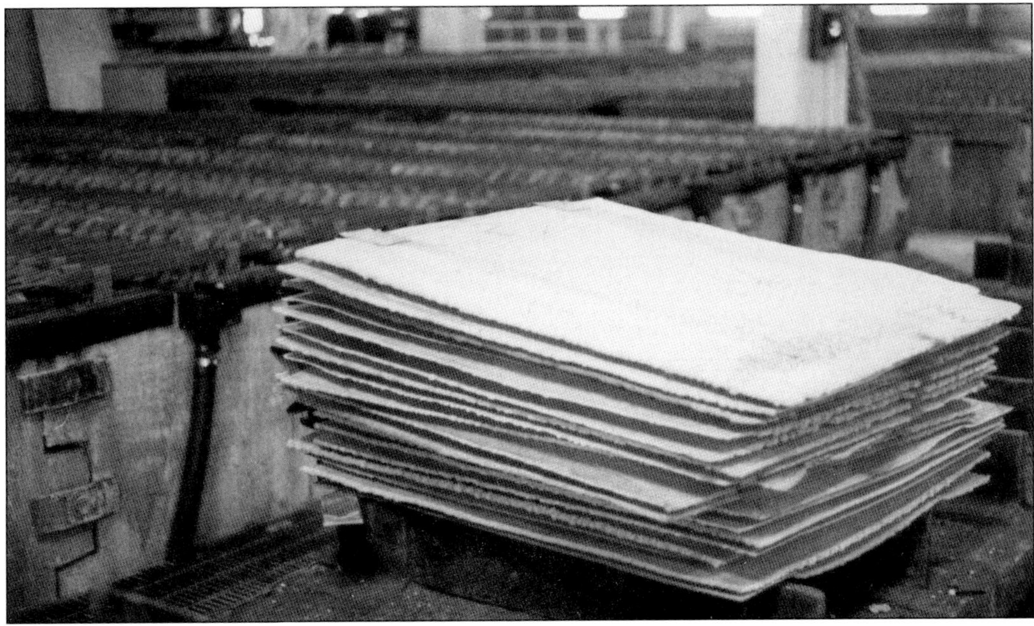

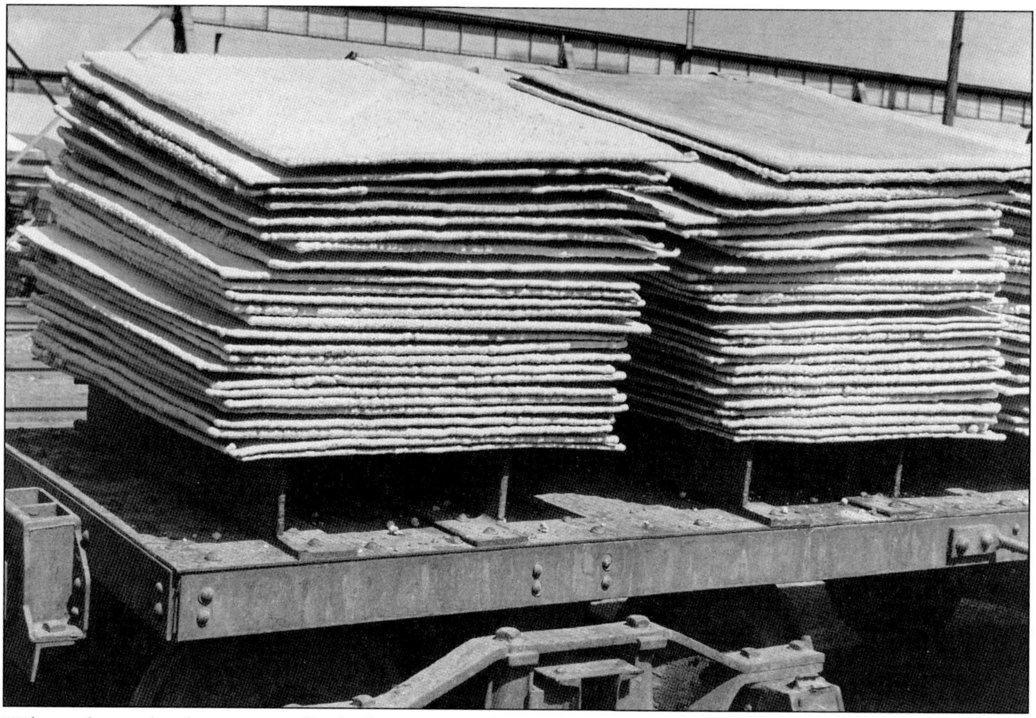

When the cathodes were pulled, they were taken to a wash tank and immersed in hot water to clean off all the electrolyte solution. The suspension bars were removed, and a cathode loader put the sheets on to tram carts, as seen above and below. The suspension bars were washed and made ready for the next set of cathodes. The electric trams moved the cathodes from the tank house to the furnace refinery. Each car carried four piles of cathodes. (Both, from 2015.089.1.)

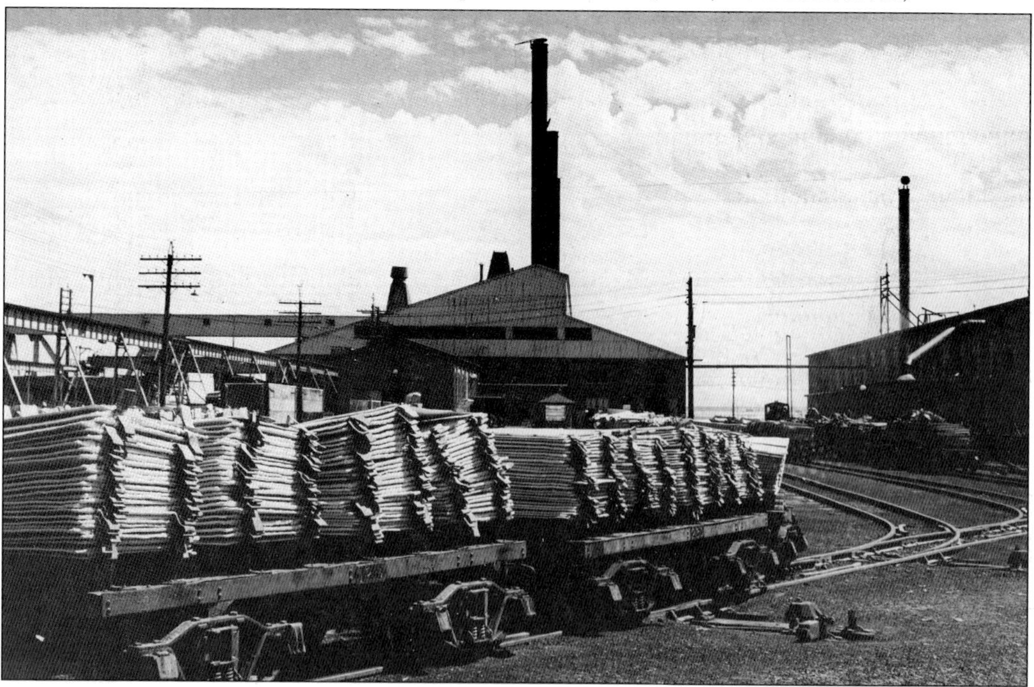

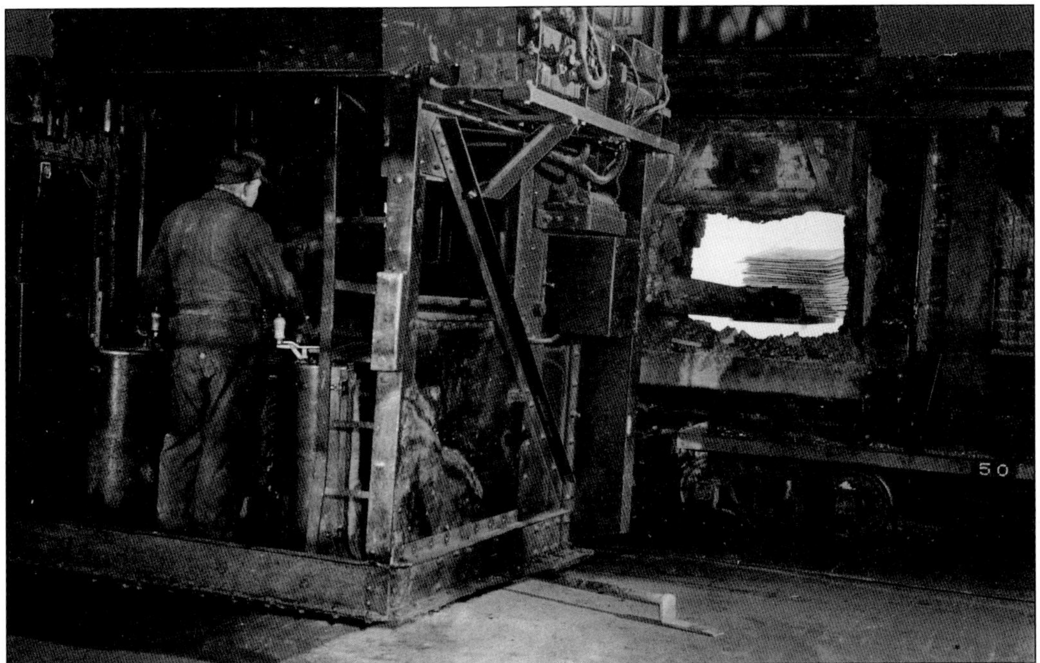

At the furnace refinery, the copper cathodes were put into reverberatory furnaces to be melted down, as seen above and below. Once the copper melted, oxygen was sprayed in to clean out other gases, called rambling, then the oxygen was removed, called poling. Green pine poles were used when poling. The cathodes were also dipped in milk of lime beforehand to provide a protective coating and reduce the amount of sulfur dioxide absorbed. The full cycle for melting and casting the copper took 24 hours. (Both, from 2015.089.1.)

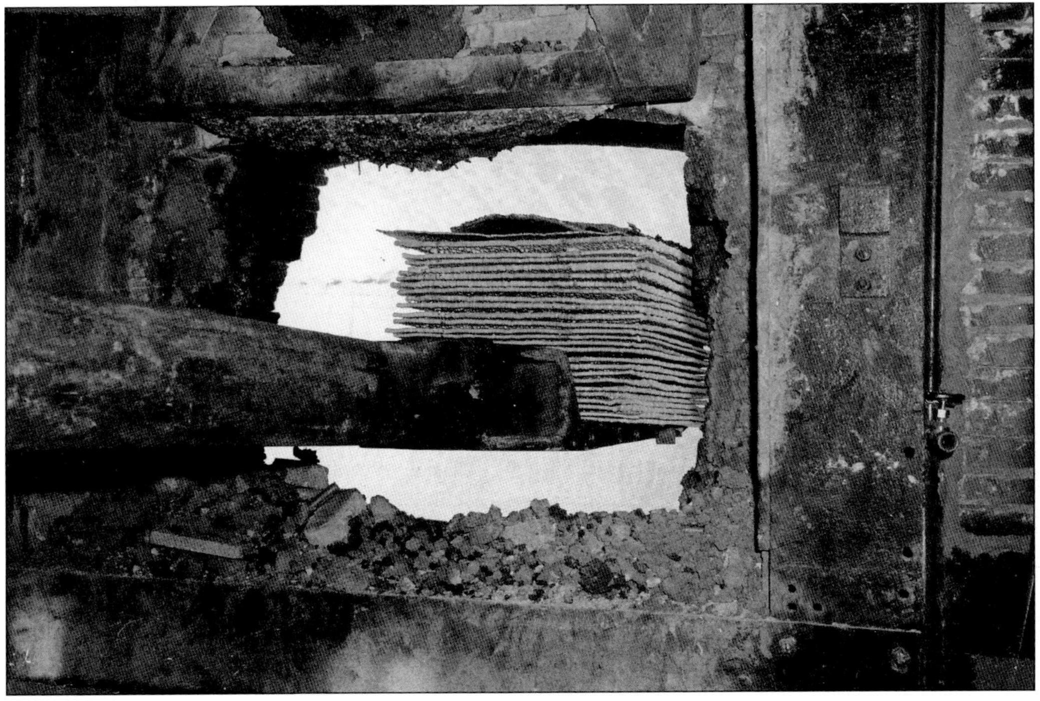

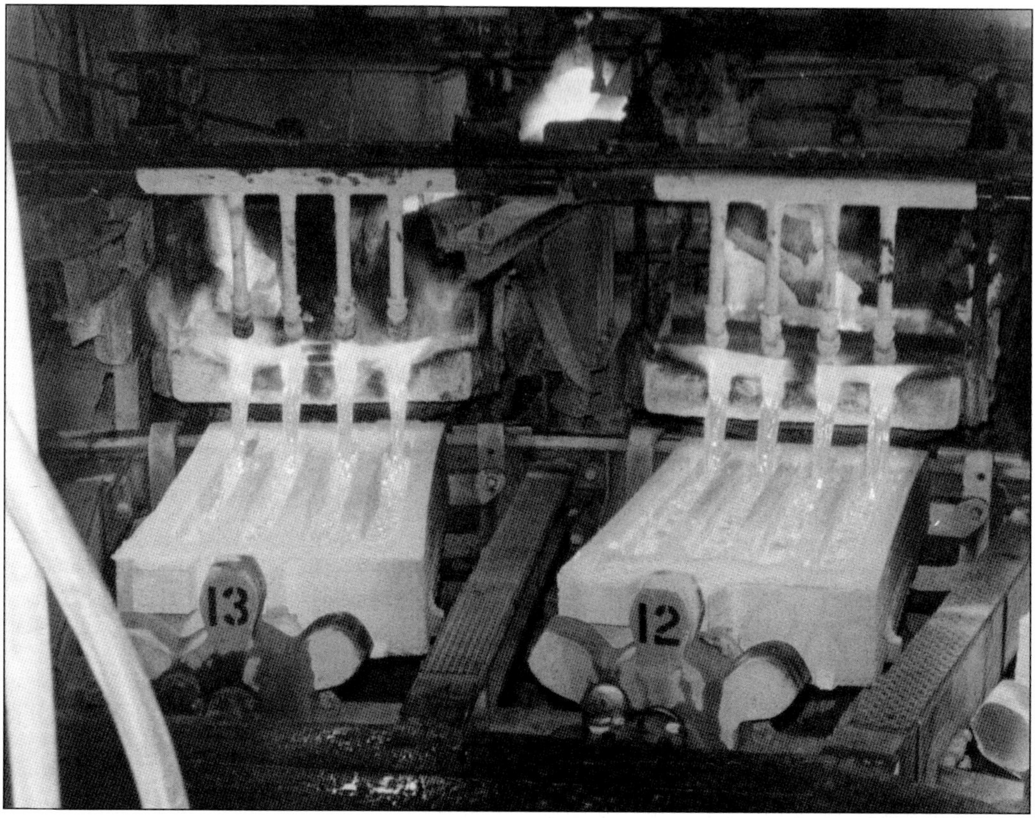

The melted copper was poured into molds and made into several shapes: wire bars, cakes, ingots, ingot bars, and in later years, round billets, as seen above. The cakes, ingots, and ingot bars were loaded into railroad cars and sent off to market. The billets went to the Billet Plant, where they were made into copper pipes. Depending on the type of casting, two to 20 castings could be gotten from each pouring of the ladle. The molds then moved along a circle and were sprayed from above and below before being dumped into a vat of water. Below is a slice from a copper bar, measuring 4.75 by 4.25 by .75 inches. (Above, 2002.085.0003; below, 1984.097.1B.)

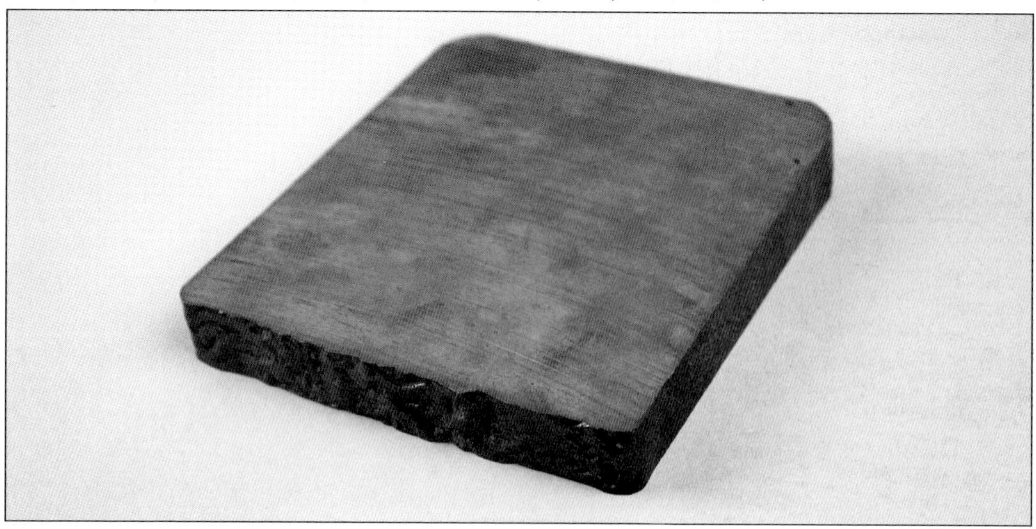

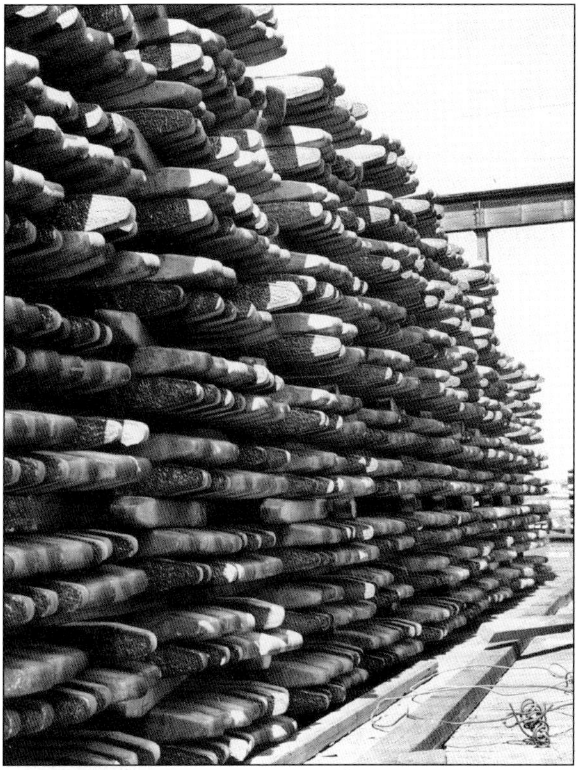

Test bars were taken from each batch of copper. Test pieces were rolled and drawn into wire and tested for conductivity and content of copper, oxygen, and impurities. Copper is known specifically for its versatility and importance in electrical work. The wire bars were five feet long and four and a half inches square and weighed 300 pounds, as seen at right. The wire bars were set aside and loaded into their own boxcar to go directly to the Wire and Cable Mill, as seen in the image below. (Both, from 2015.089.1.)

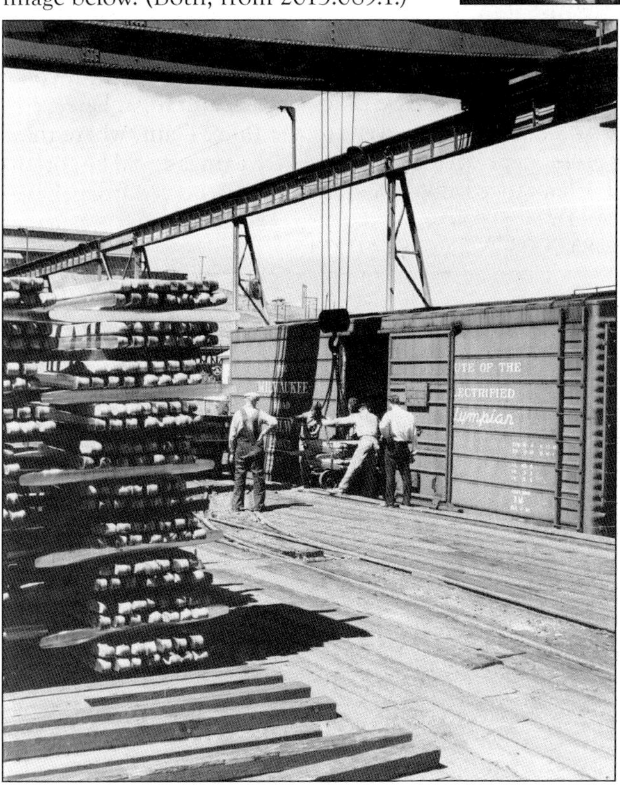

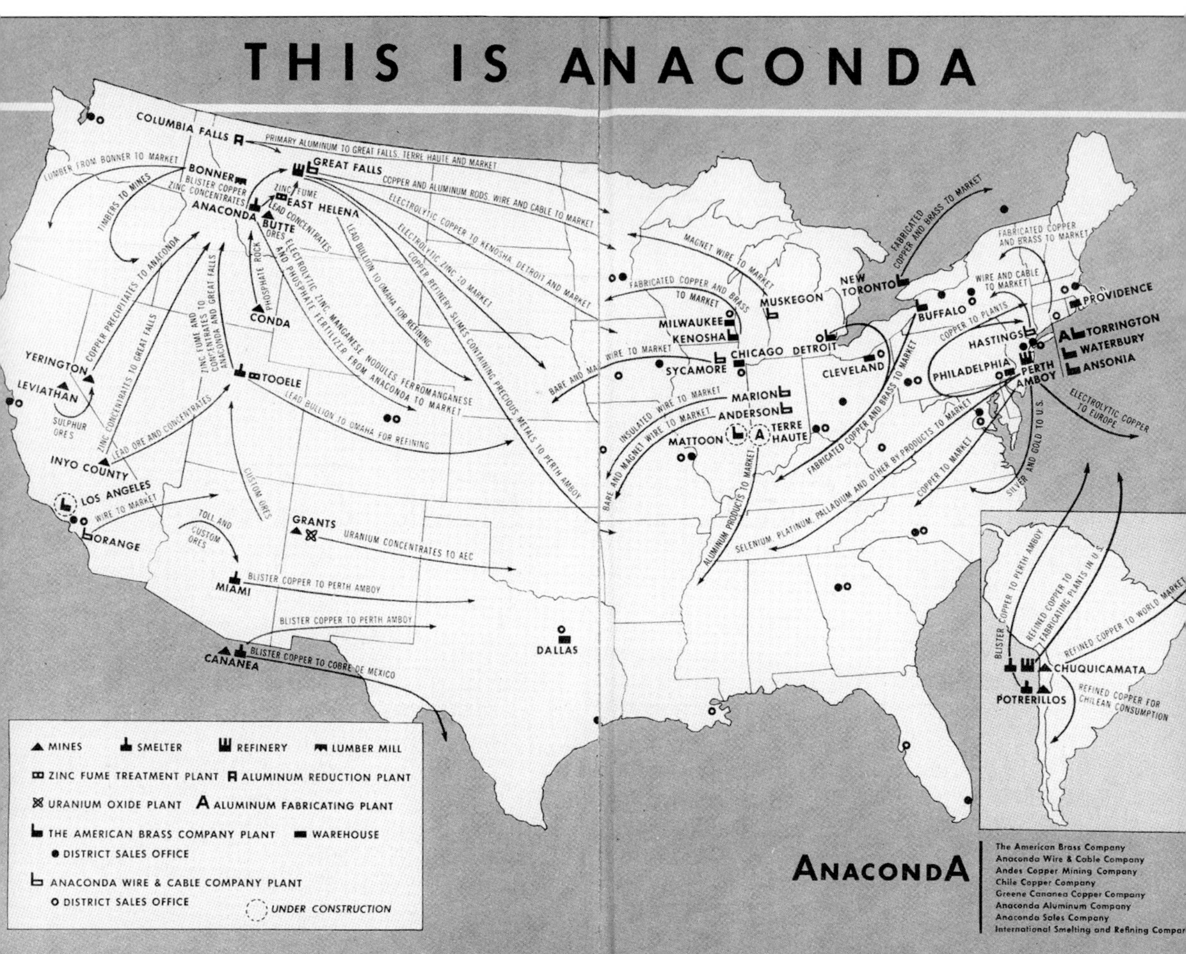

The map shown above gives a much better understanding of Anaconda's reach and distribution at the height of production and processes. Great Falls was a key part of the line of production as the only refinery in the West for the Anaconda Company. The company's reach was worldwide not only in distribution but in acquisitions as well. (From *Anaconda* by Isaac F. Marcosson.)

Three

THE PROCESS FOR ZINC

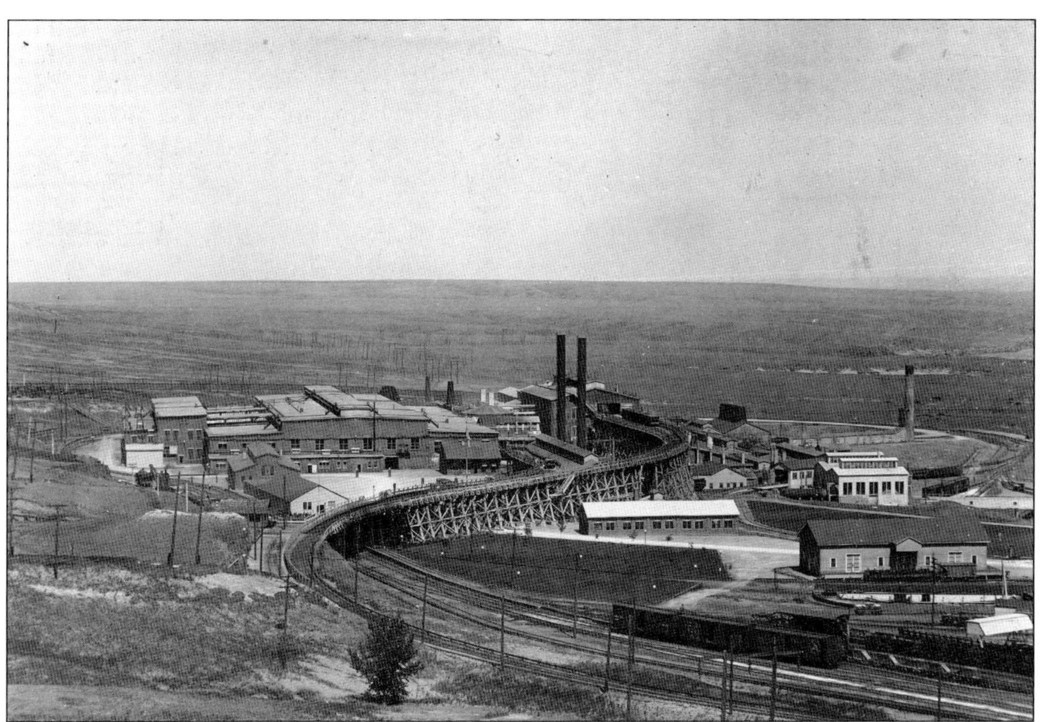

In 1916, while the copper refinery was updated, the smelter added a zinc processing plant, shown above, capable of producing zinc using a commercial electrolytic process. Originally, the electrolytic process could only be used on complex lead-silver zinc ores unsuited to the traditional pyrometallurgical processes. More development made it possible to apply the electrolytic process to any type of zinc ore. By 1970, the plant had expanded its capacity to 500 tons a day. (From 2015.089.1.)

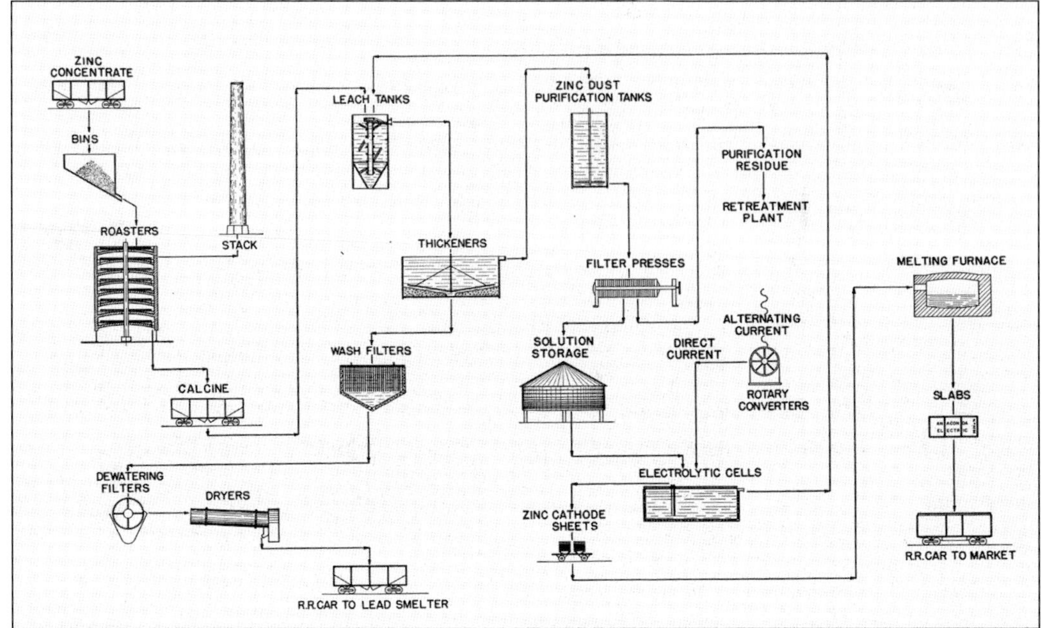

The electrolytic zinc process separates operations into three divisions: preparation of the concentrate for leaching, treatment of roasted concentrate to produce a solution of zinc salt to be electrolyzed, and the electrolysis of the solution to produce metallic zinc and regenerate the leaching solution. Although seemingly simple, the diagram above illustrates the complicated nature of the process. The plant was expanded in 1926 to meet demand, as seen below. (Above, 1991.119.0026; below, 2008.099.0030.)

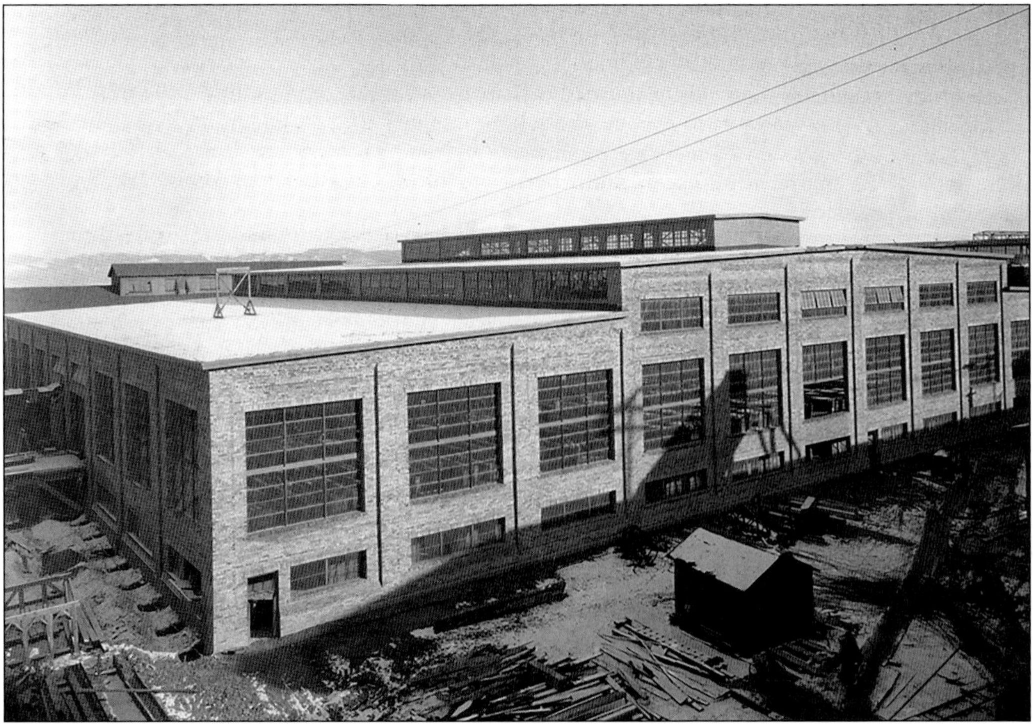

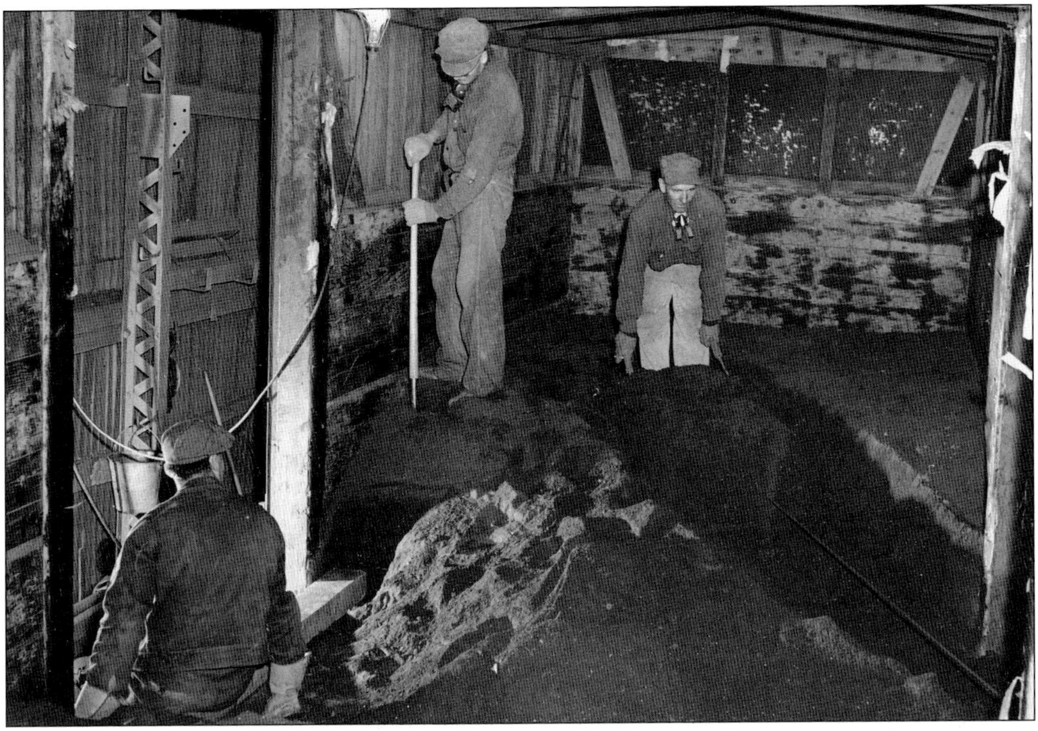

In the beginning, Butte and Anaconda were the main sources for zinc concentrate, but over time, more and more came from global sources. The zinc content in the concentrates varied widely, between 43 and 57 percent, and they included such elements as copper, lead, gold, silver, cadmium, iron, cobalt, germanium, indium, aluminum, manganese, magnesium, silica, tellurium, selenium, gallium, and sulfur. All of these had to be removed to purify the zinc. Over time, improvements in concentration practice resulted in higher-grade concentrate. Zinc processing began with zinc concentrate arriving at the plant in railroad cars that were unloaded into under-the-track bins for immediate use in the roasters or storage. From left to right, Joe Koefelda, Frank Bates, and Walter Koenig are seen above unloading zinc concentrate into hopper bins. Below, Bob Koleff (left) and Ed Jette (right) are seen unloading excess zinc concentrate into storage for later. (Above, 1982.017.0355; below, 1982.017.0267.)

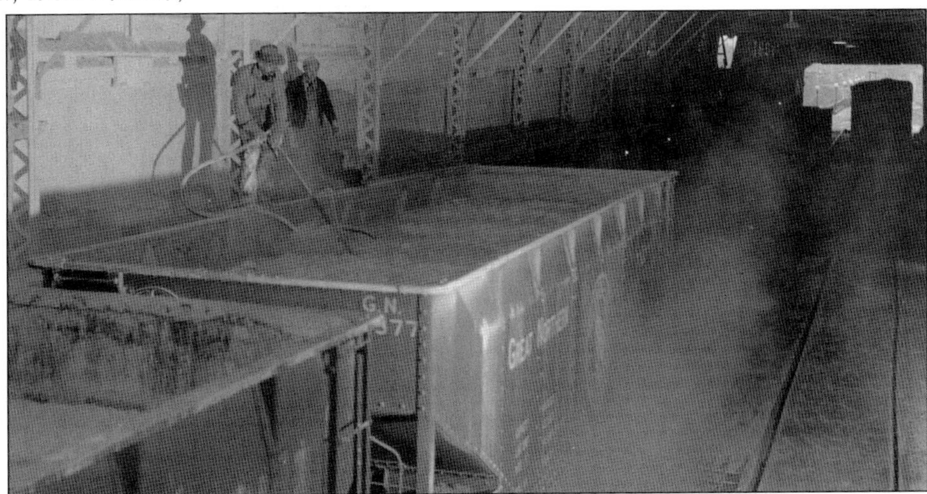

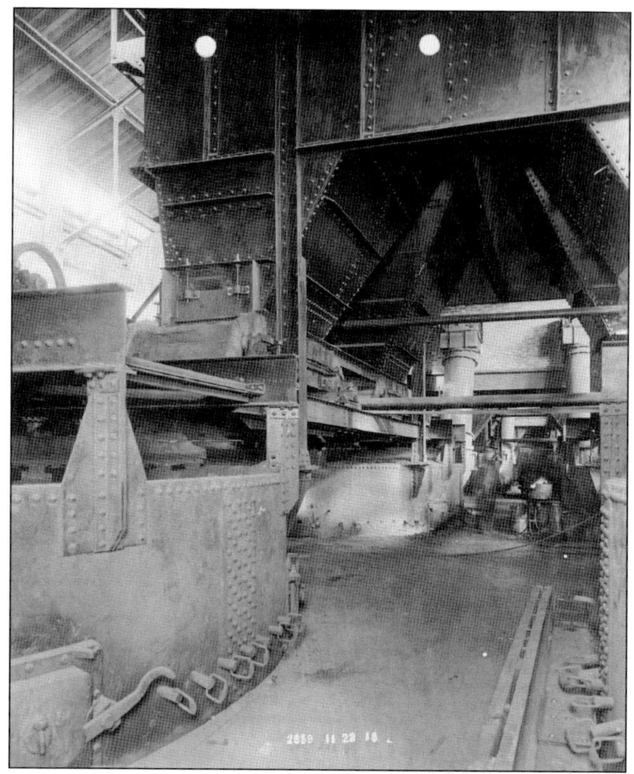

Zinc concentrate arrived as zinc sulfide, which is not soluble and must be roasted to become readily soluble zinc oxide. Roasting is the most important part of the electrolytic process, specifically when it includes leaching with diluted sulfuric acid. How well the leaching works decides how much of the metal is recovered in the rest of the process. Below, Joe Slanina (left) and Walter Bodeen (right) of the Roasting Division are seen moving concentrate from the storage bins to transport to the roasters. They charge the hoppers, at left, that feed the roasting furnaces to keep production going. (Left, 1982.017.0211; below, from 2015.089.2.)

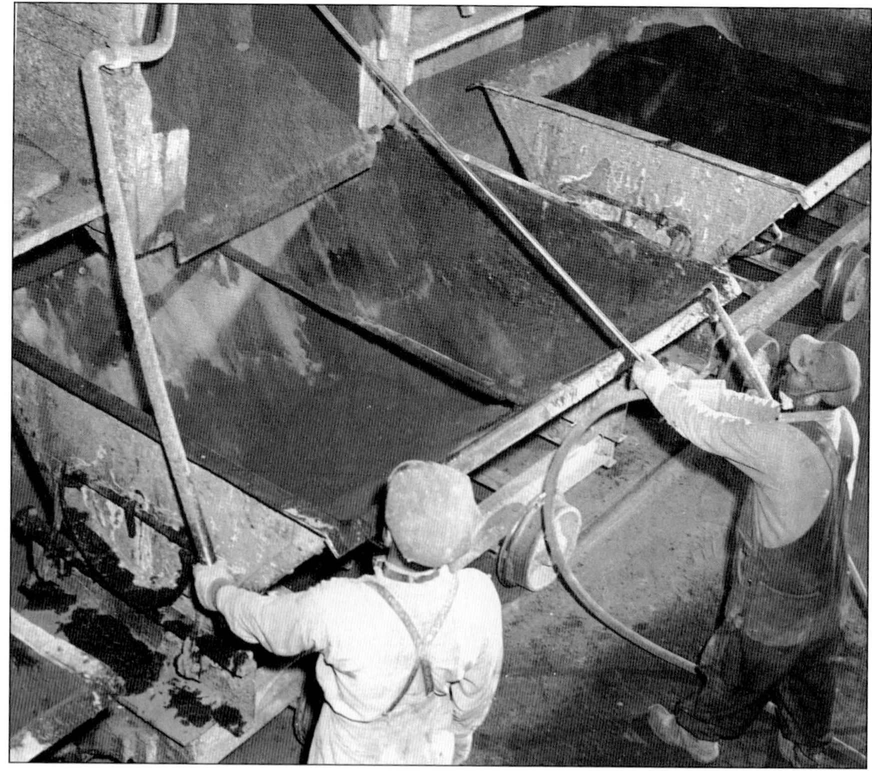

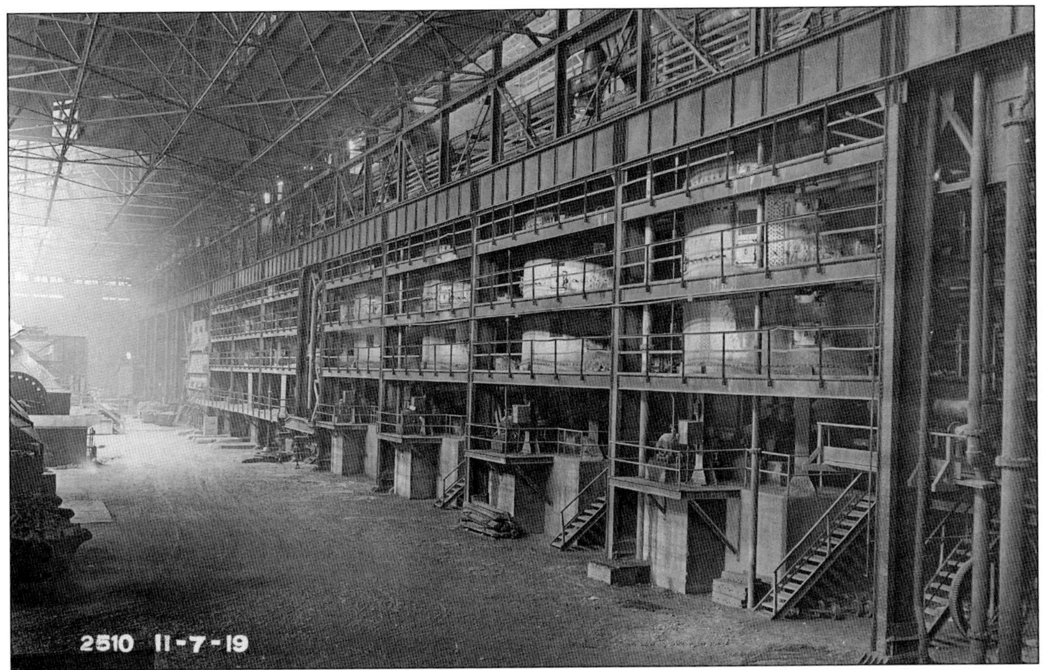

Wedge furnaces were used for roasting the concentrate, as seen above. They were circular with a steel- and brick-lined exterior and a hollow center occupied by a revolving shaft in the middle with 26 cast-iron arms. Roasting depends almost entirely upon the control of temperature and time. The roasted material, called calcine, fell through holes in the arms all the way down to the seventh hearth, then into a hopper. The hoppers discharged into cars to move to the leaching department. George Grecich is seen below doing just that. (Above, 1982.017.0130; below, 1982.017.0264.)

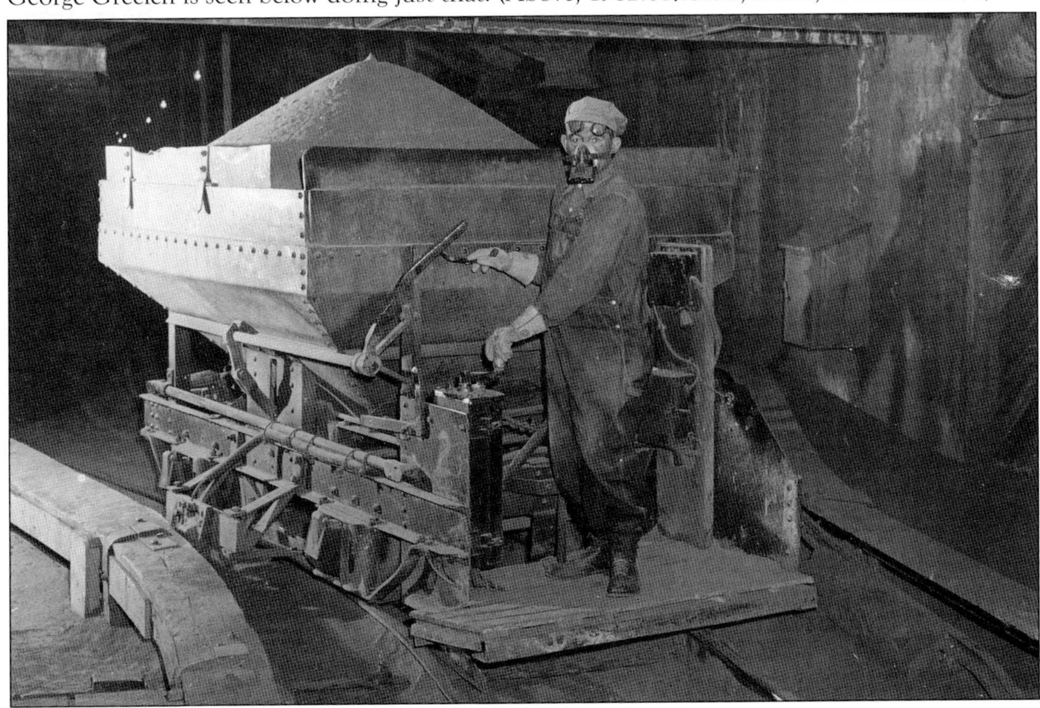

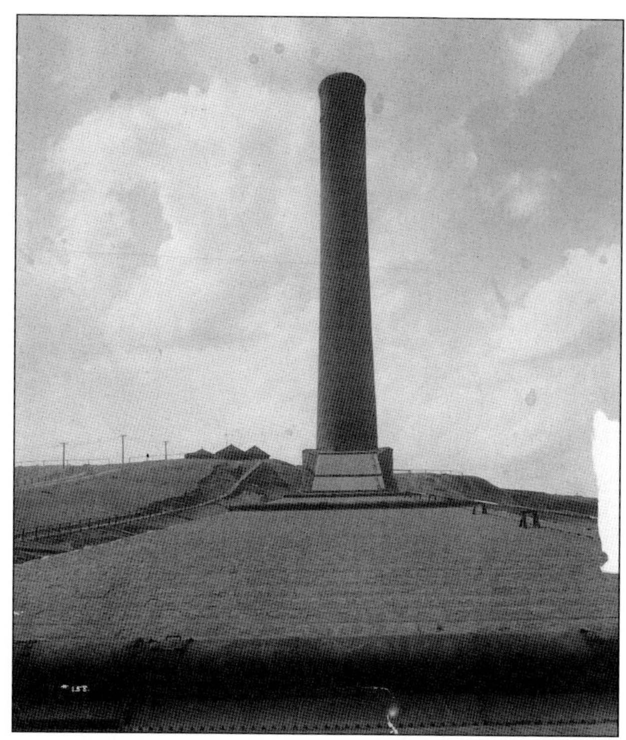

In the roasting steps, the sulfuric gases were passed into offtake flues that led to the main gas flue and the smokestack, seen at left. Dust collected from the roasting furnace flues was retreated along with the oversized material that had migrated to the bottom of the furnaces. The furnaces had to be cleaned every day or they would become too hot and result in failed chemical results. Charlie Larson of the Roasting Department is seen below cleaning a furnace. (Left, 1990.026.0034; below, 1982.017.0268.)

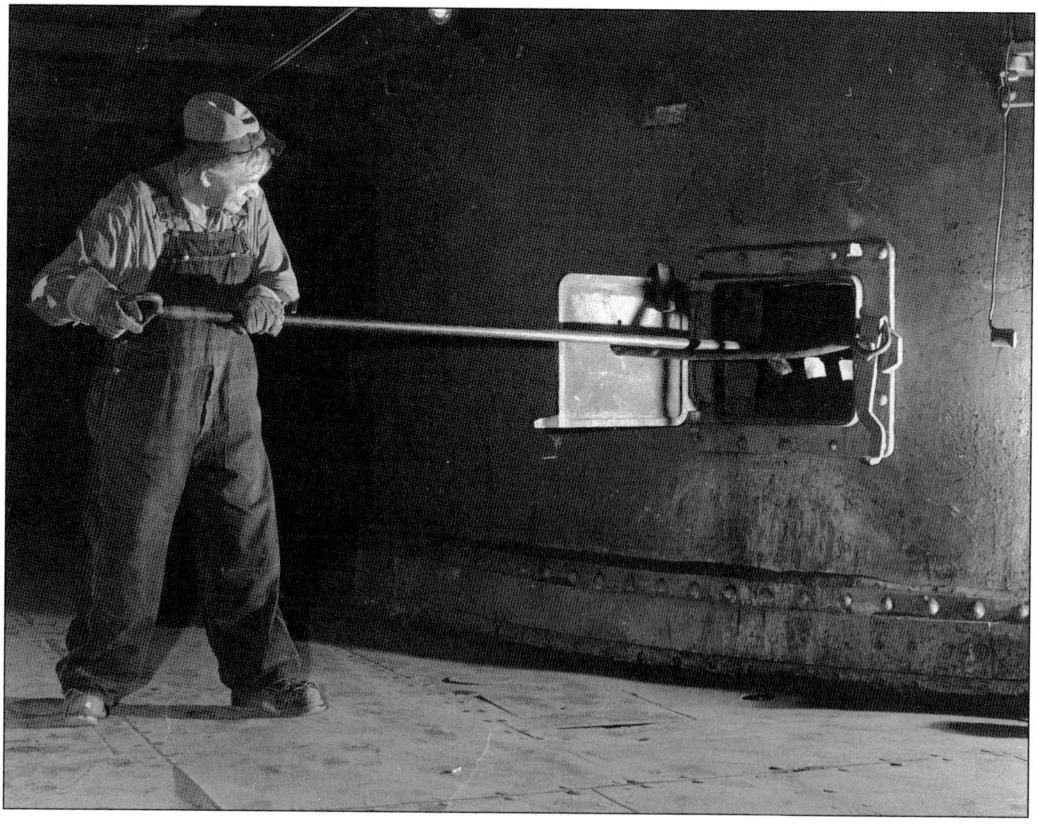

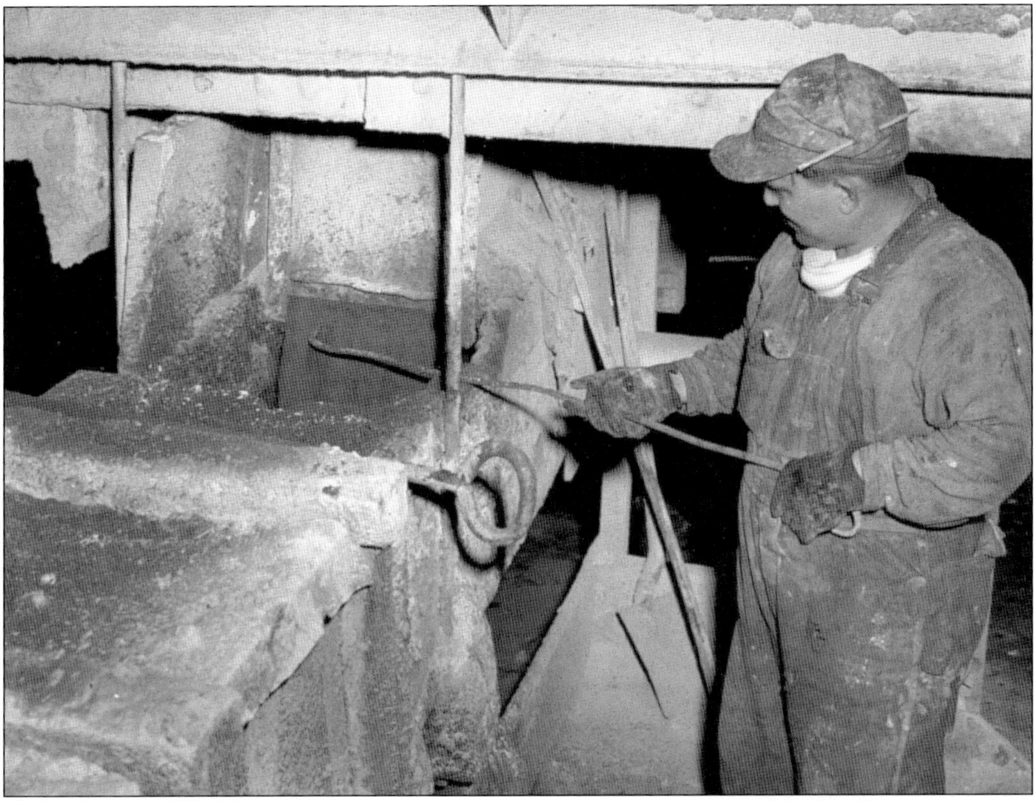

Roasted calcine was dissolved in sulfuric acid, and the metal was obtained in a metallic state through the electrolysis of the solution in a process called leaching, which also purified the solution of unwanted elements. Purification of the solution is necessary because average zinc ore contains many other elements besides zinc. In the first cycle, or neutral leach, the calcine was continuously introduced into a stream of return acid from the electrolyzing process. Loren Gewald is seen above adding the calcine. The neutral leach ran continuously through several tanks, pictured at right, to agitate the solution with air. (Above, 2023.043.0170; right, 2023.043.0111.)

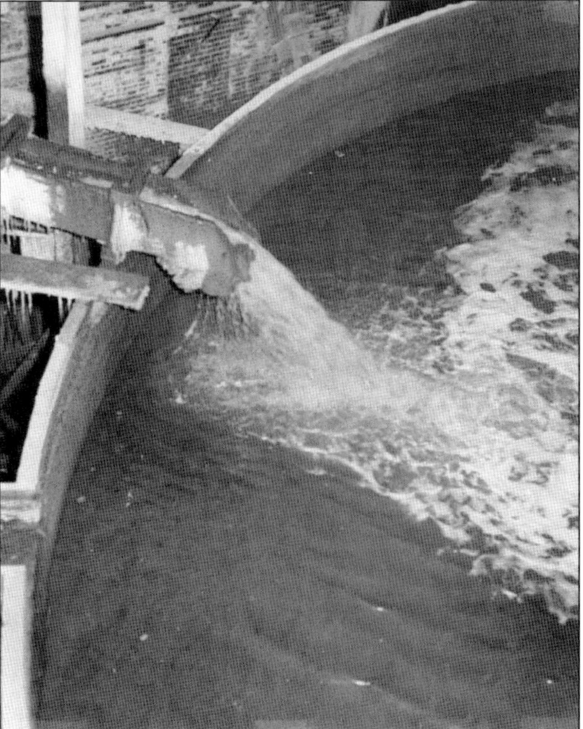

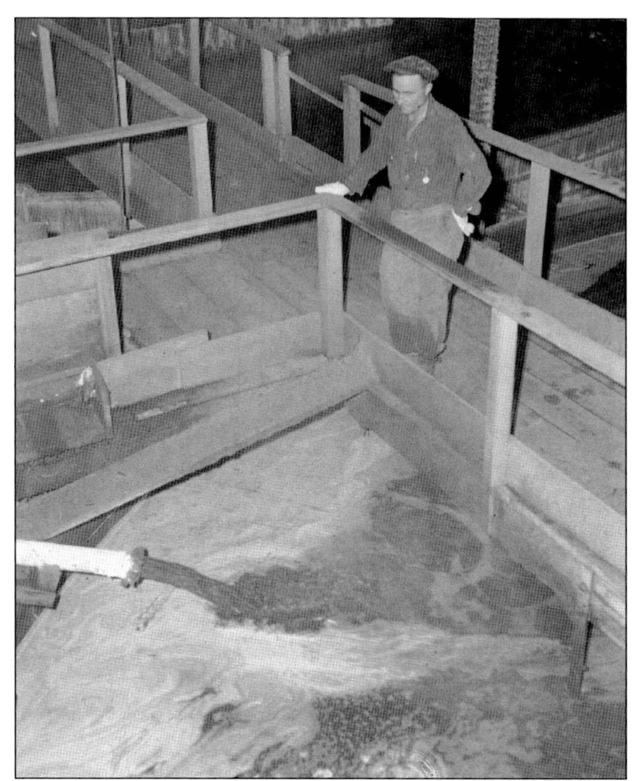

The solution was fed into large settling tanks called Dorr thickeners; Edward Farrell watches in the photograph at left. The massive tanks, 50 feet in diameter and 12 feet high, could hold 750 tons of solution, and it was there that the solids were separated. The clear solution moved to an overflow tank, where it was further purified by adding zinc dust to the solution, which was then agitated through a series of purification tanks. Below, John "Sandy" Sanderson is adding the zinc dust. (Both, from 2015.089.2.)

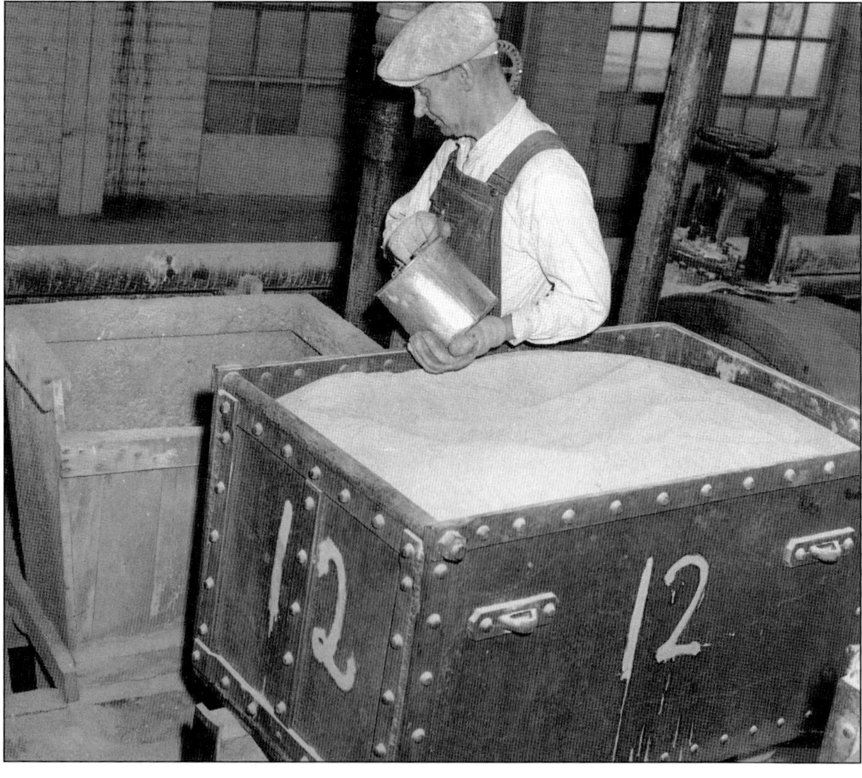

The zinc dust caused the unwanted elements to be deposited as sludge. The solution was then sent through a special Schriver press, where the sludge was removed. Carl Stimac turns on the press at right. Zinc could be produced from this process alone, but it would not be economically practical. Gay L. Holbrook is shown adjusting a spigot on the Schriver press below. (Right, from 2015.089.1; below, from 2015.089.2.)

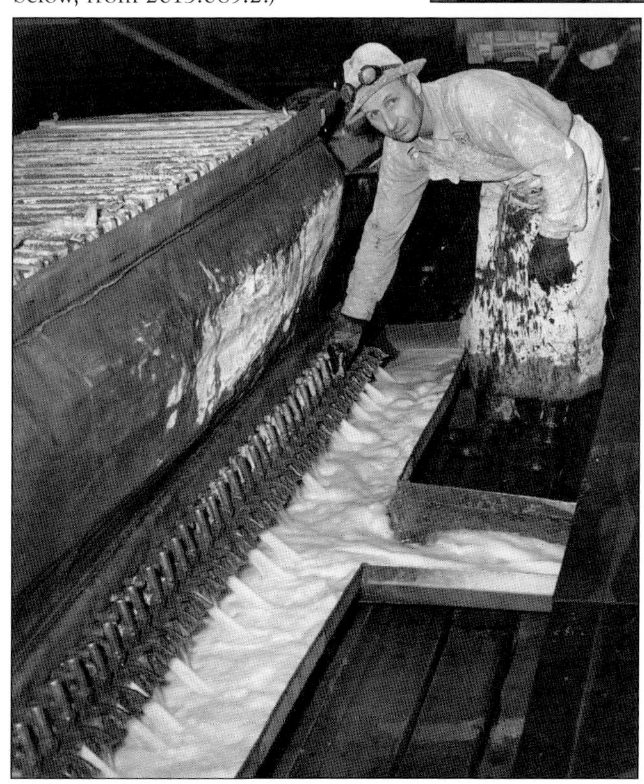

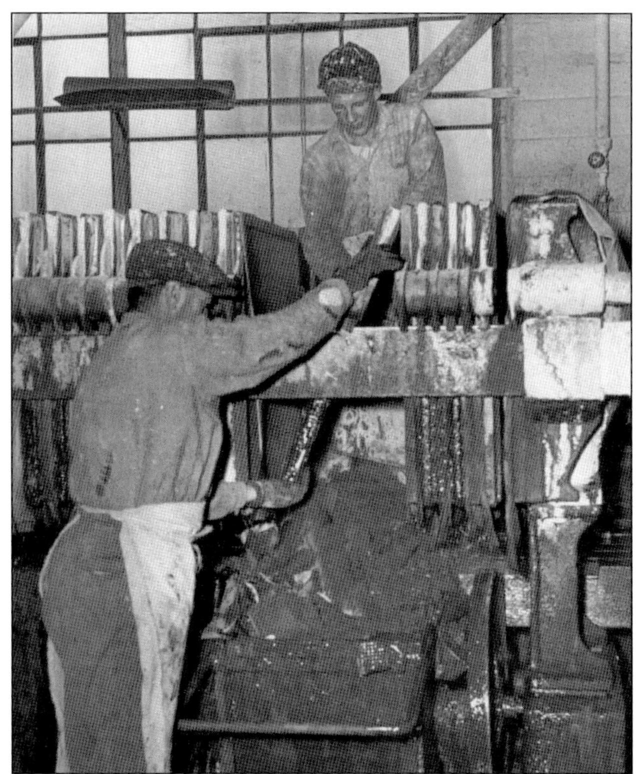

To make the process profitable, the residue from the Schriver press and the solids recovered from the Dorr thickeners must be retreated, as there was still a high percentage of zinc to be extracted. At left, Harold Britt (left) and Bill Linn (right) are seen removing the residue from the Schriver presses. There were three more cycles to go through to get the most zinc possible. The residue from the Dorr thickeners, along with some sandy materials previously extracted from the neutral leach discharge, made up the feed for the second cycle, or acid leach. A classifier, seen below, separates the sandy material from the first cycle residue and feeds it into the appropriate tanks. (Left, 2023.043.0059; below, 2023.043.0090.)

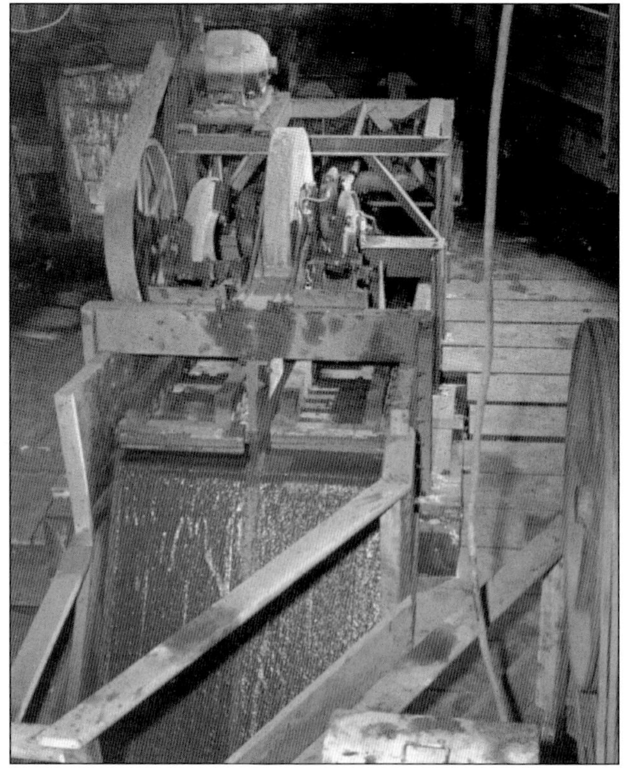

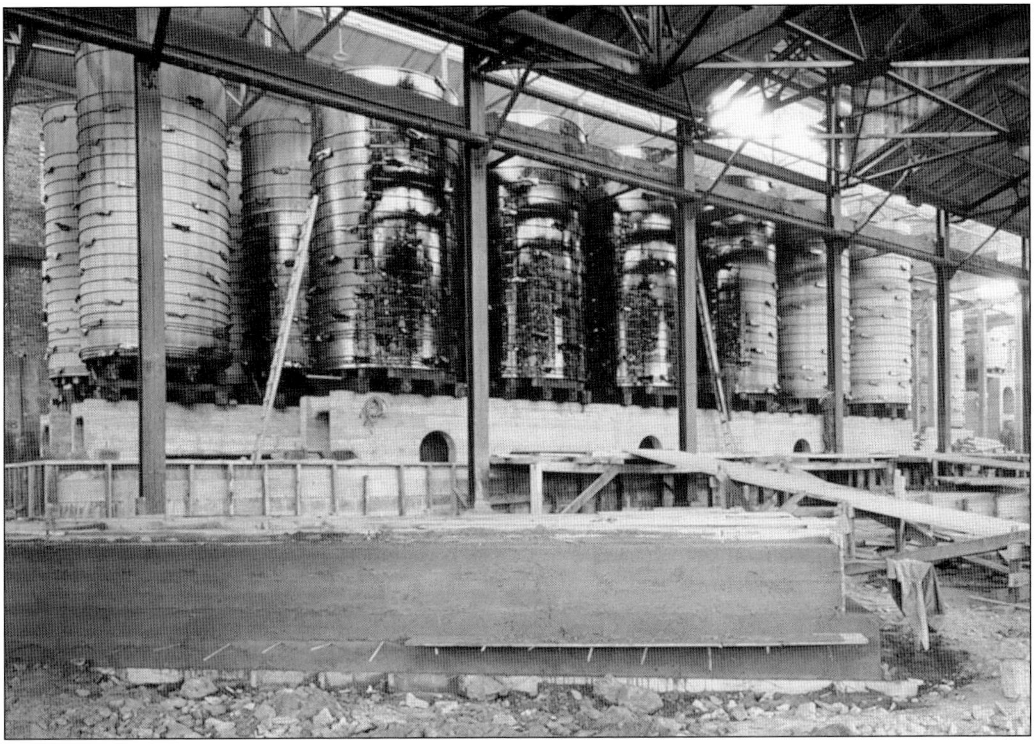

The acid leach went through a series of eight Pachuca tanks, where the solids from the first leach and the returning acid from the Electrolyzing Department were mixed with more dissolved zinc. Above, they are being built on November 3, 1927. Once again, the solution was sent to the Dorr thickeners, and the solids settled to the bottom. This time, however, the solids were moved from the bottom of the thickeners and carried to revolving drum-type Oliver filters, seen below. There, all the excess solution was removed from the solids. (Above, 2008.099.0005; below, 1982.017.0206.)

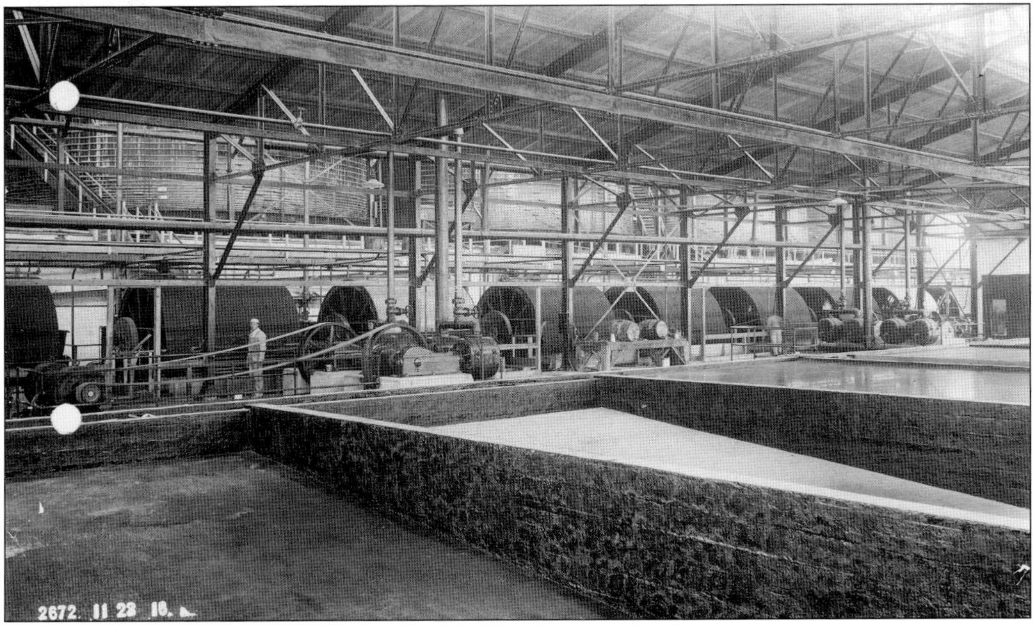

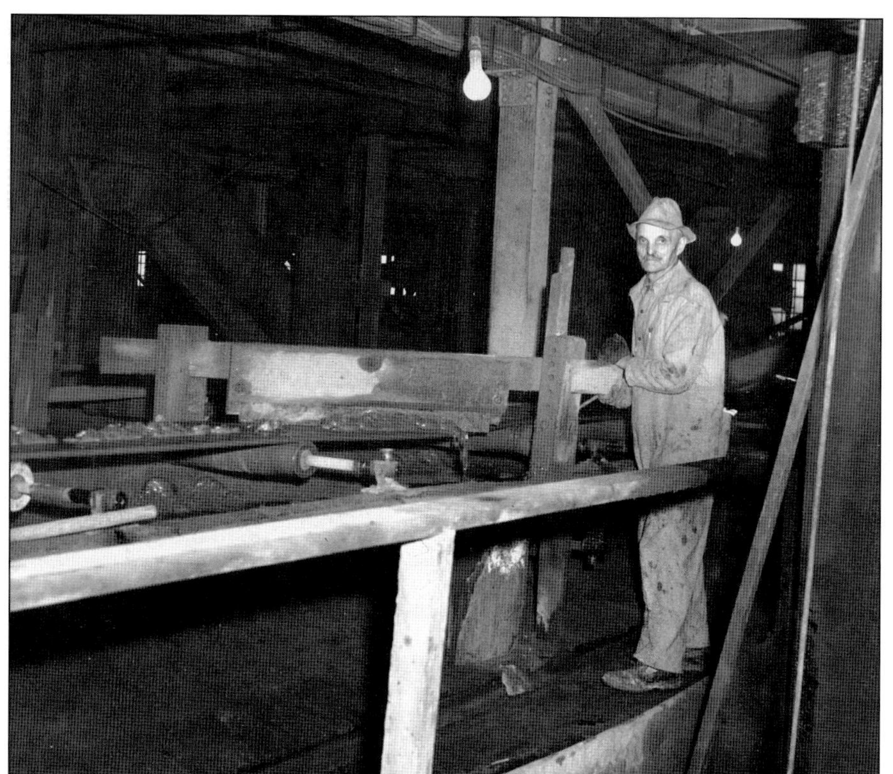

The residue was carried by means of a conveyor belt to the beginning of the third cycle, the residue leach. Conrad Jorgenson is seen above adjusting a scraper on the conveyor belt. Eighteen mechanically agitated leach tanks were used to divert the residue into the designated residue leach tank. Higher acid strengths were used to dissolve the zinc from the solids. For a final time, the solution was sent to the Dorr thickeners, and the settled solids were further thickened and washed in a leaf-type Moore filter. At right, John Moy is raising a filter for cleaning. The remaining solids were dried and put into railroad cars to be shipped to the Slag Treatment Plant in East Helena. (Above, from 2015.089.1; right, from 2015.089.2.)

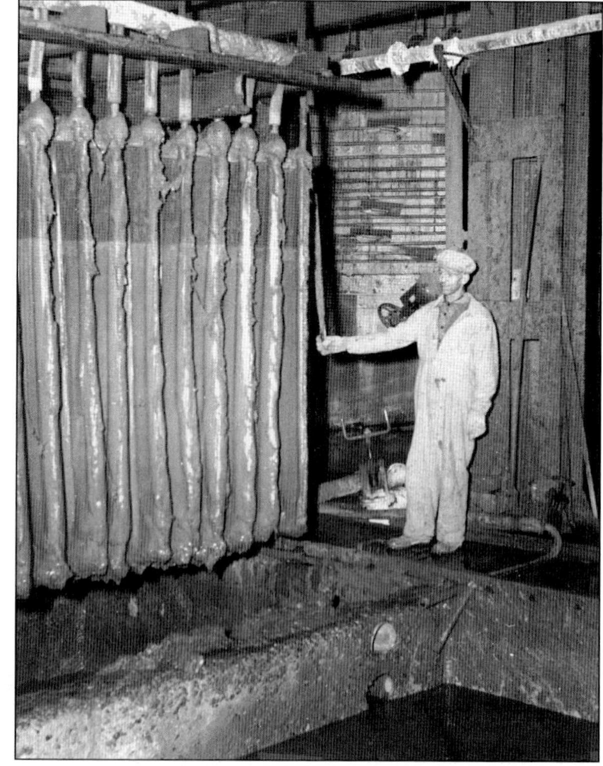

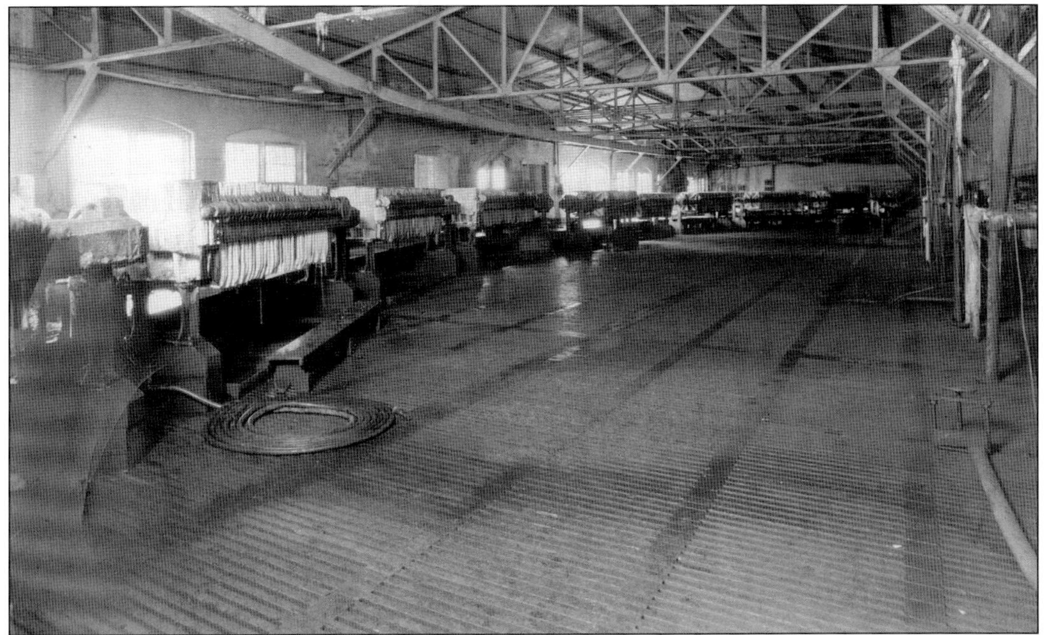

For the fourth cycle, the residue from the Schriver presses was oxidized by air drying and then leached with the return acid from the electrolyzing cells. Above is the interior of the old Schriver building. The residue returned to the Dorr thickeners, where it was treated with more zinc dust to push out the copper and cadmium that were still in the residue. At this point, the copper was sent to the copper smelter in Anaconda, and the cadmium was sent to the Cadmium Building on the local grounds. The solution went back to cycle one. Samples were taken frequently from several points in the Leaching Plant, as Jim Lind is doing below. (Above, from 2015.089.1; below, from 2015.089.2.)

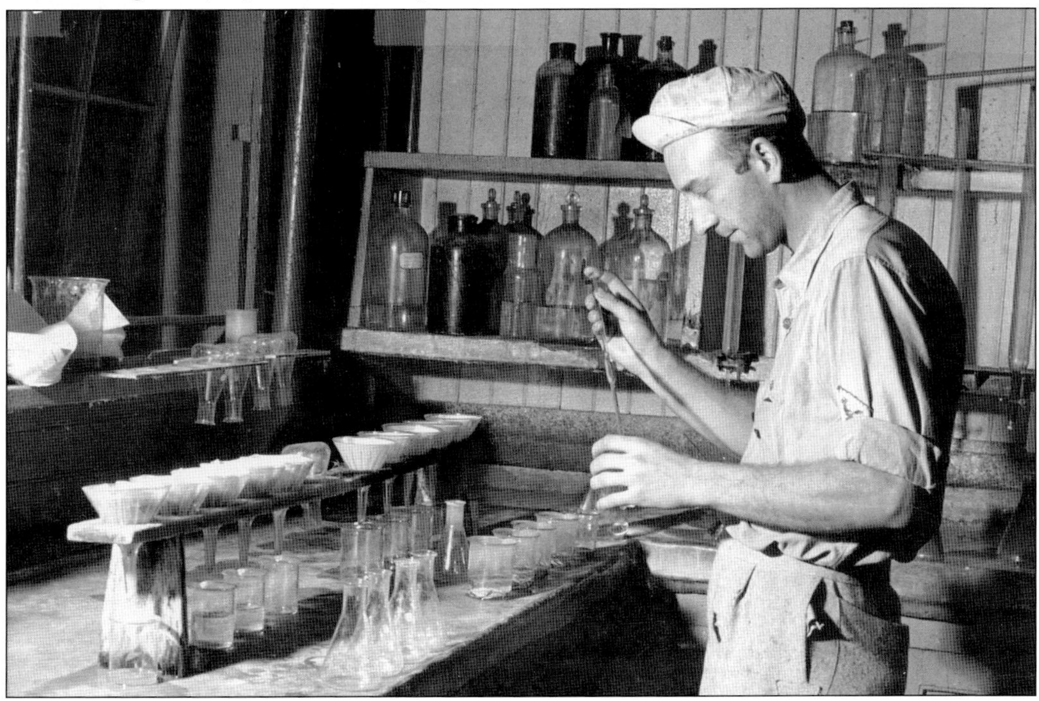

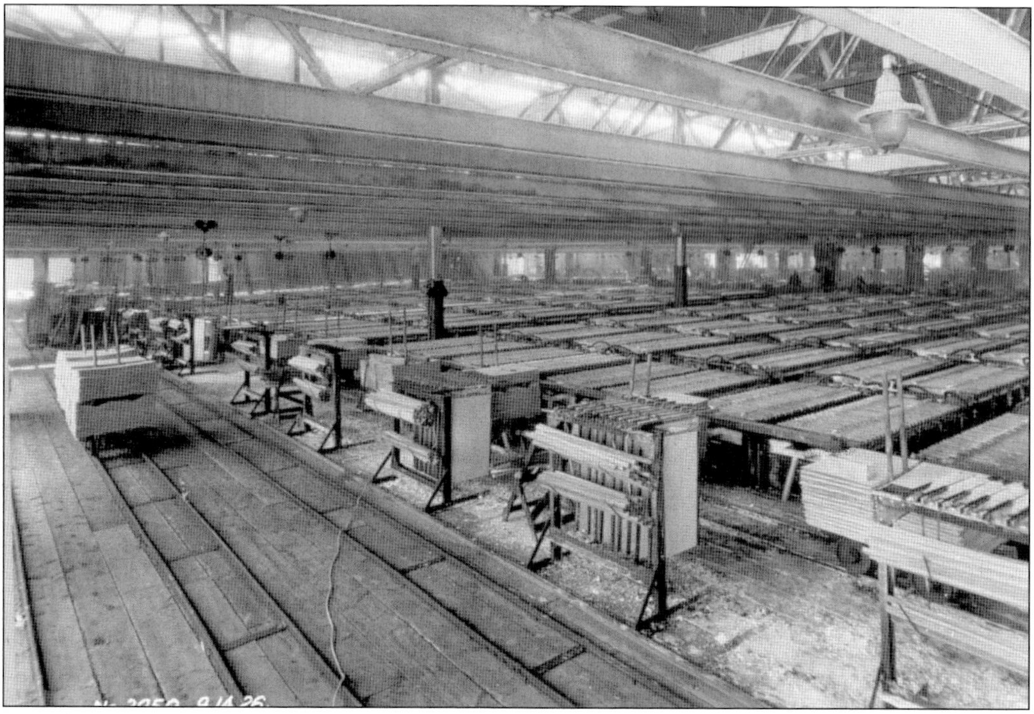

Now that the leaching process was completed, the purified zinc sulfate solution went to the Electrolytic Building. The cells were lead-lined wood and concrete tanks, as seen in the tank room above. The anodes were made of lead, and the cathodes were made of aluminum, with copper header bars for support. Rubber-lined strips were placed along the vertical edges to help with stripping the zinc sheets and to eliminate sharp edges, as seen below. (Above, 2023.043.0011; below, 2008.099.0043.)

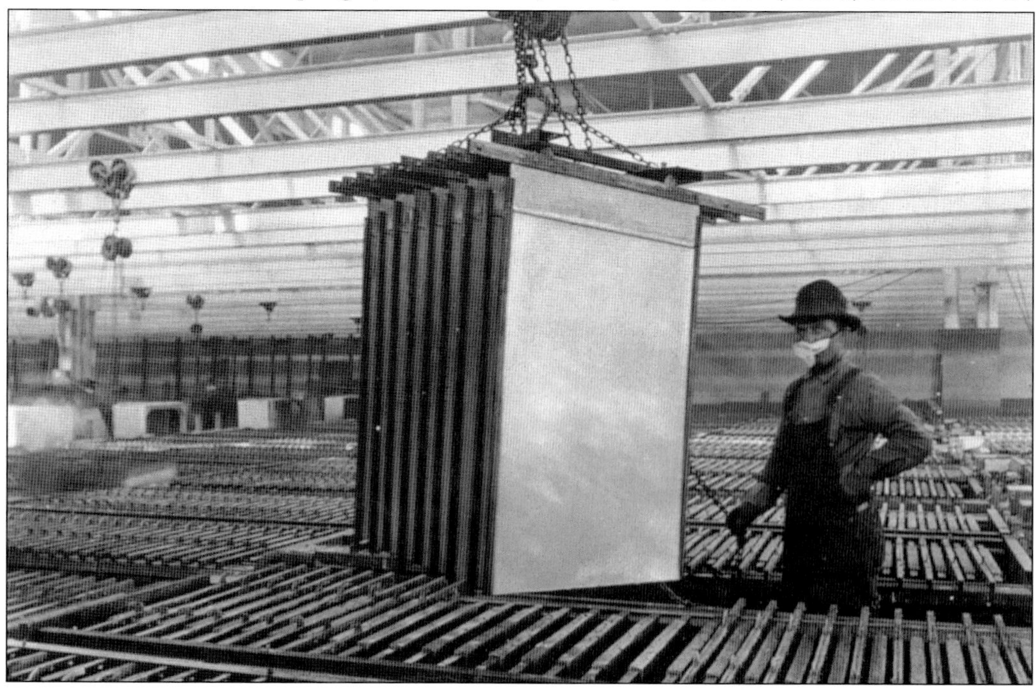

There were eight units consisting of 144 cells that were arranged so the solution cascaded through a series of six cells. Each cell held 27 anodes and 28 cathodes interleaved. On the aisle side of the cells were copper bus bars to carry the electrical current. The fresh solution from the leaching department had no free acids, but as the zinc attached to the anodes, acid was formed. The used solution was returned to the leaching division to be recharged. At right, Ernie Bowga (left back) and Art Louma (right front) are shown in the zinc electrolytic cell room, and below is a broad view of the same room. (Right, from 2015.089.2; below, from 2015.089.1.)

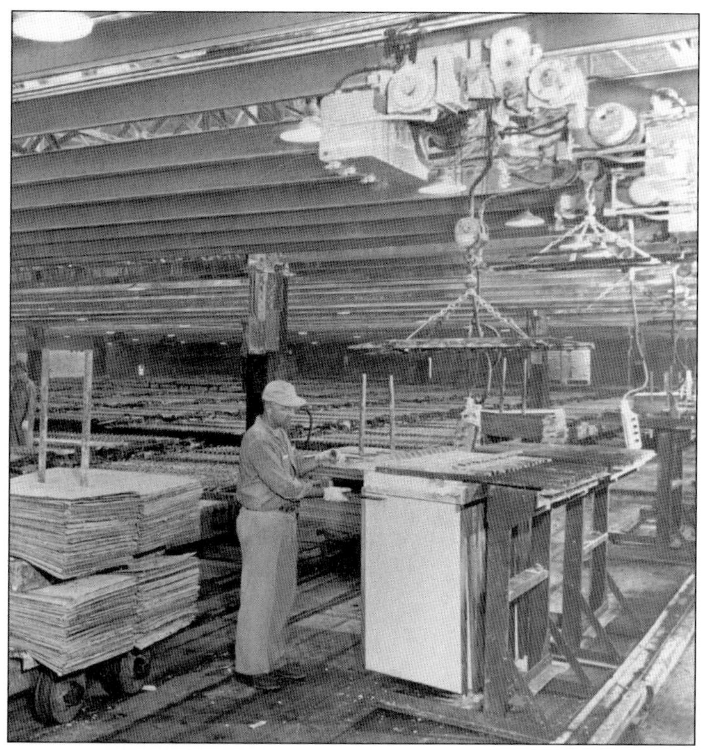

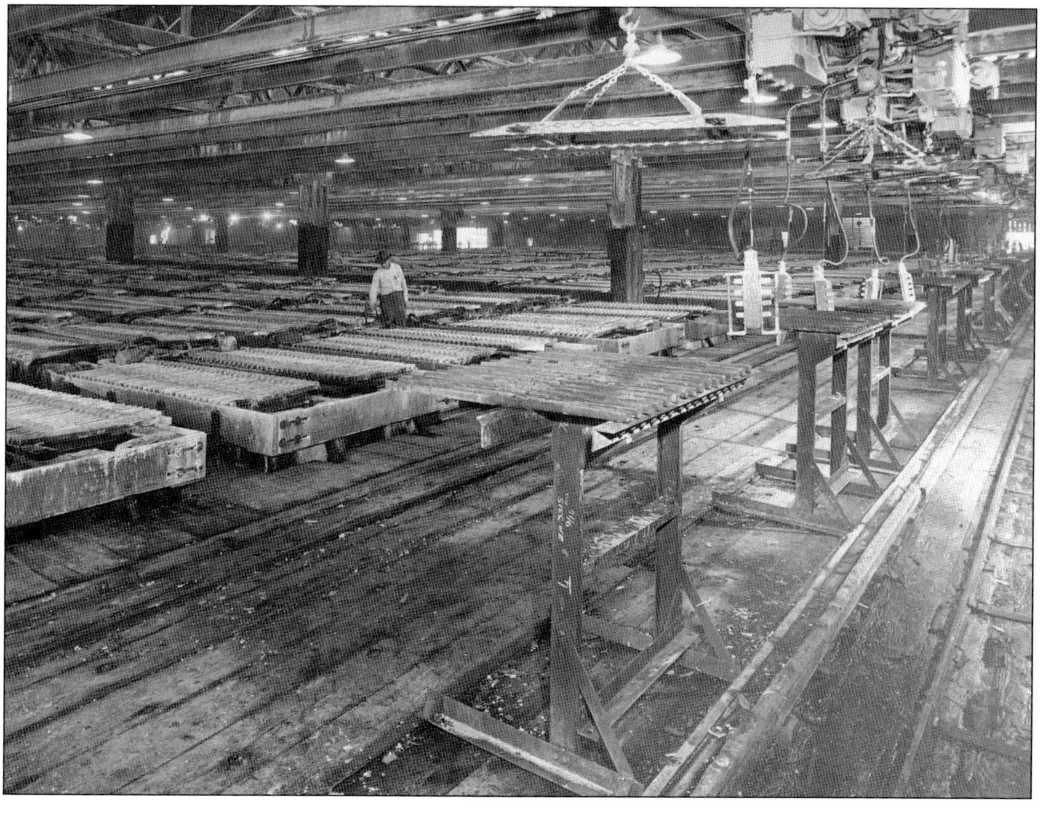

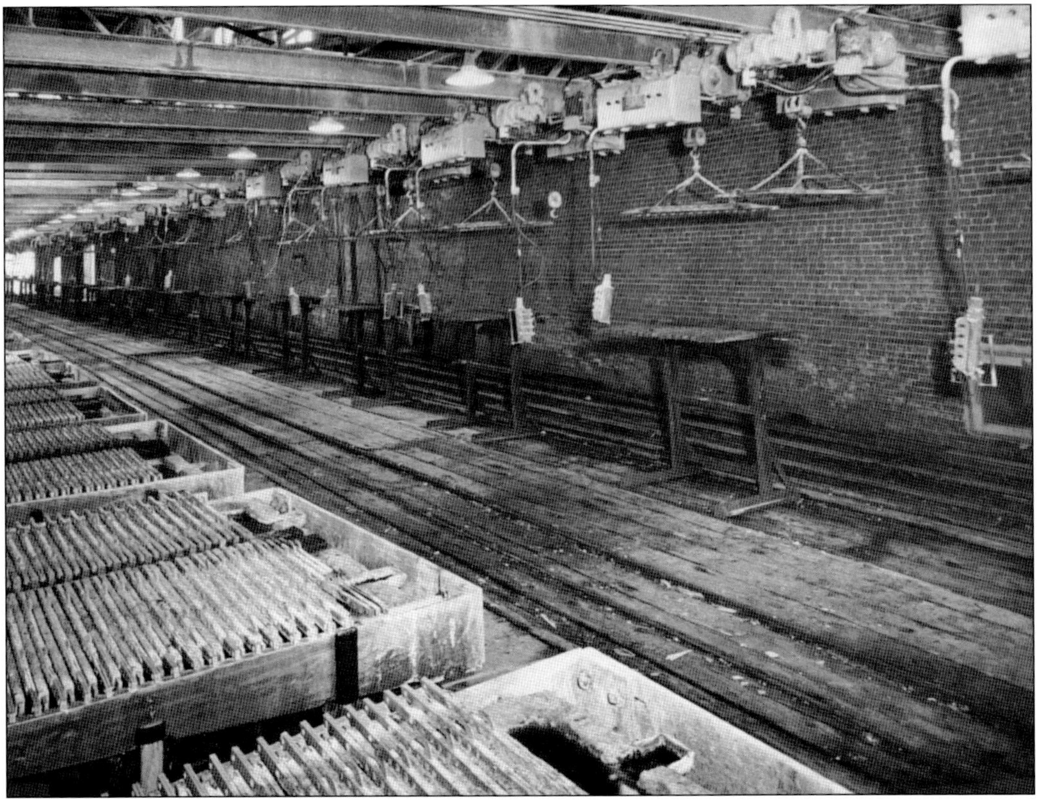

The electrical current caused the zinc sulfate to break up into its component parts: zinc, sulfur, and oxygen. Zinc was deposited on the cathodes in metallic form. The oxygen became gas, and sulfur combined with water to become sulfuric acid. Every 24 hours, the zinc was stripped and put into piles to be transported to the casting division. Above is a stripping station within the zinc electrolytic cell room, and below, Harvey Wallett pulls zinc sheets. (Above, 2023.043.0275; below, from 2015.089.2.)

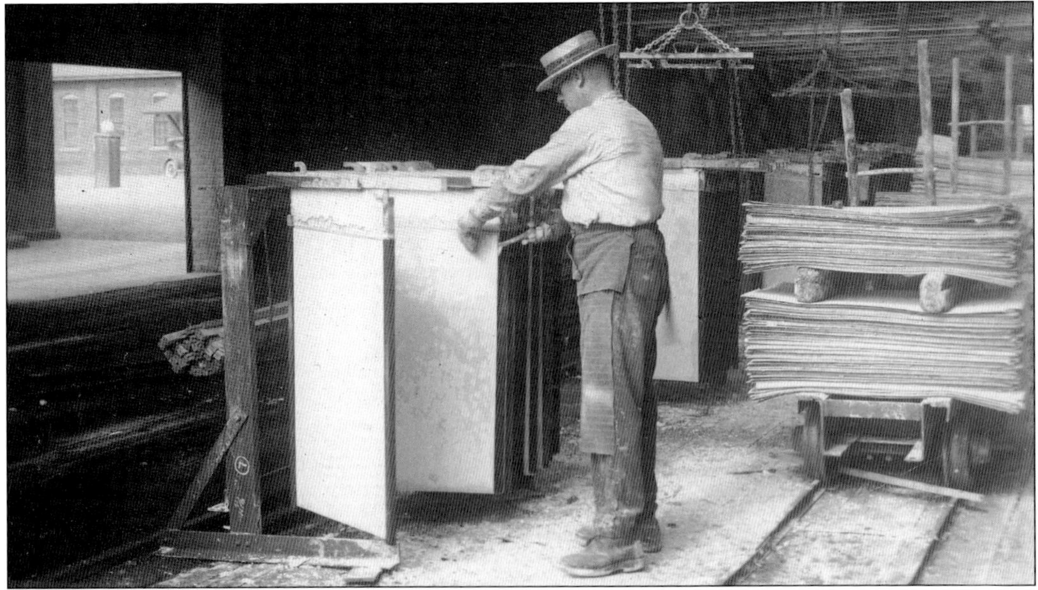

The zinc sheets were stacked in straight piles weighing 2,500 pounds each. The entire bundle was put into the furnace and became immersed in molten metal. At right, Joe McElroy (left) and Charles Kilmas (right) are charging or loading a melting furnace with zinc sheets. Oxygen in the air reacted with the zinc and caused dross (zinc oxide) to form. Long-handled hoes were used to skim the furnaces. The dross would otherwise act as an insulation agent and hinder the furnace's ability to melt the metal. Below is the No. Three Furnace. (Both, from 2015.089.1.)

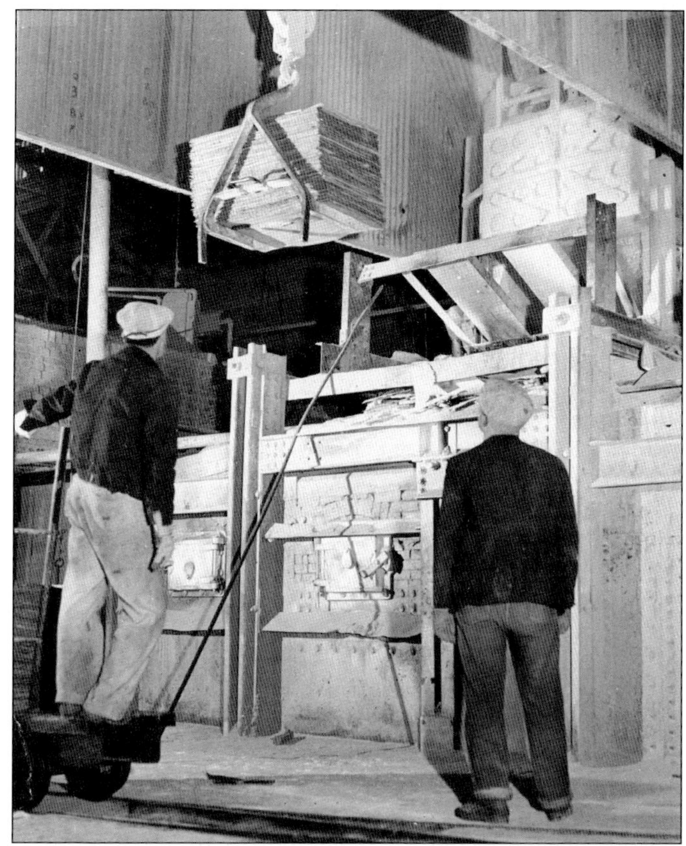

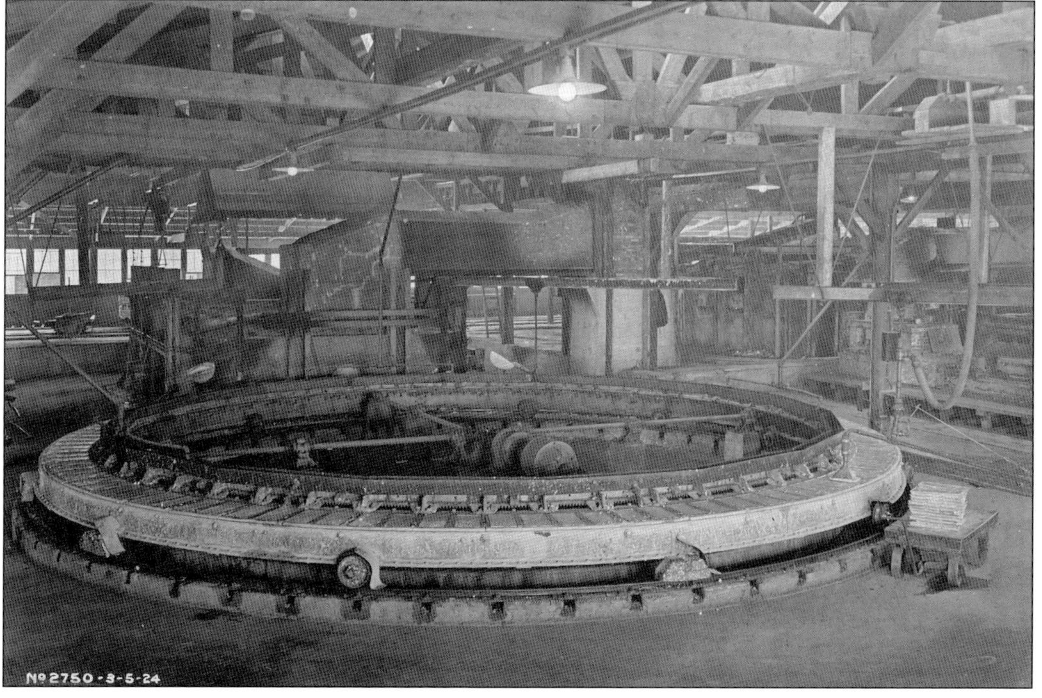

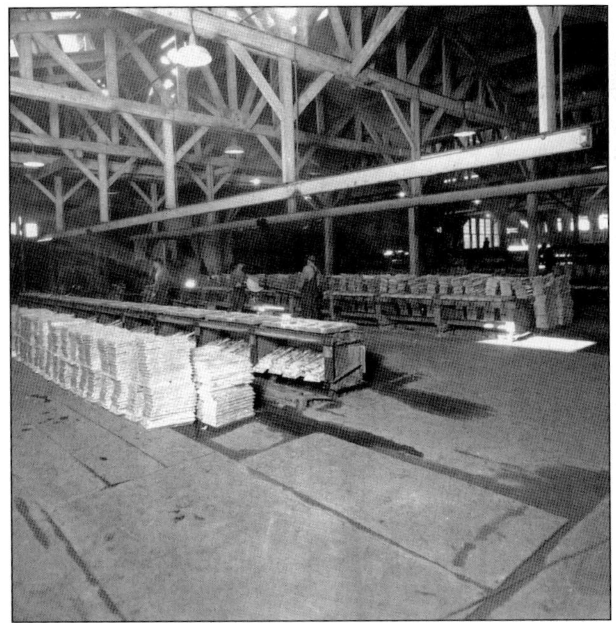

The Great Falls Zinc Plant was a custom smelter, and the zinc metal was cast for four or five toll customers in addition to the Anaconda Sales Company. Toll customers provided their own raw materials and had the plant process it for a fee. Approximately 55 percent of total production was for the Anaconda Sales Company. Besides the various slab molds, there were 10 different types of cake molds ranging in size from 500 pounds to 5,000 pounds. There were four grades of metal produced as well: special high grade, high grade, prime western, and alloy. At left is what zinc casting looked like before 1930, and below are Robert Smith (left) and Alvin Nash (right) in the Zinc Casting Building. (Left, from 2015.089.1; below, from 2015.089.2.)

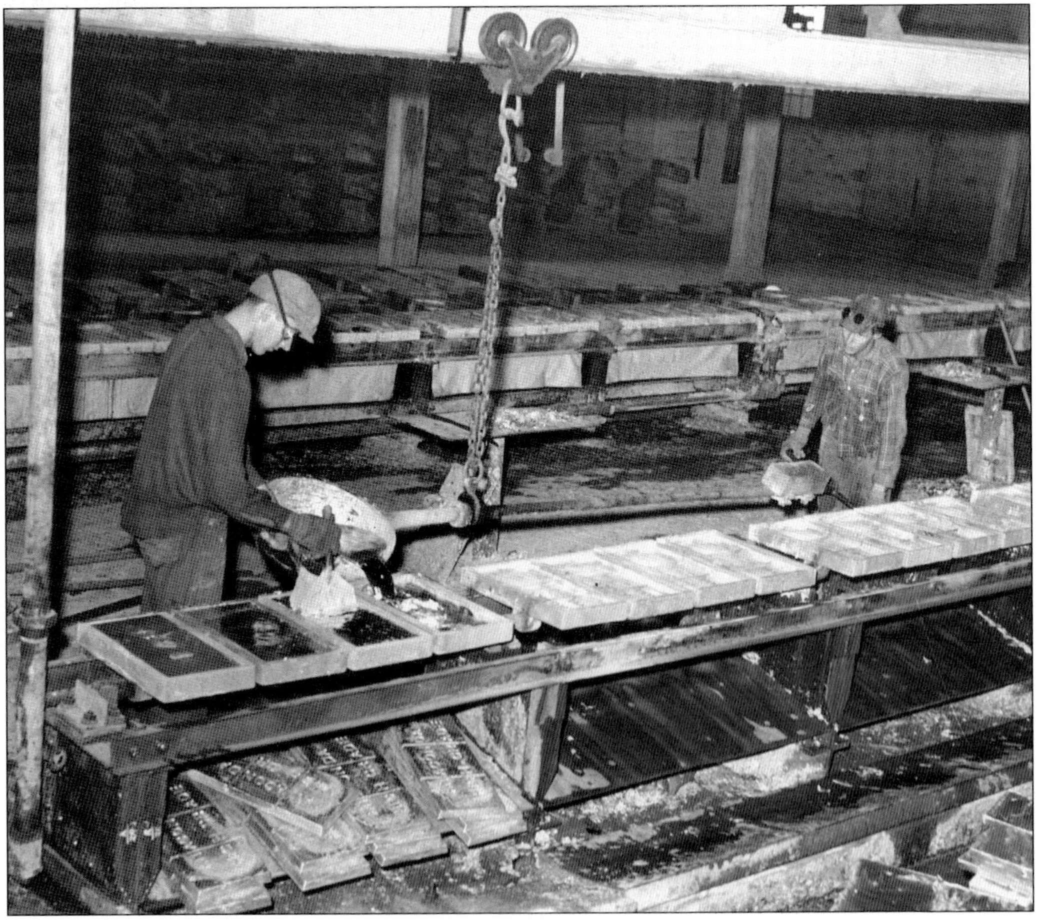

To produce special high-grade metal, the leaching needed to eliminate the cadmium in the purification step. The No. One and Four Furnaces were used for high-grade metal. Depending on monthly toll customers and Anaconda sales requirements, the No. Two Furnace was used for both high grade and special high grade. Most alloys were produced in the No. Three Furnace. Ladles were dipped into the molten metal and carefully poured into a line of molds, one at a time. At right, Robert Smith is filling a zinc mold, and below is a closer look at the molds. (Right, from 2015.089.2; below, 1982.017.0304.)

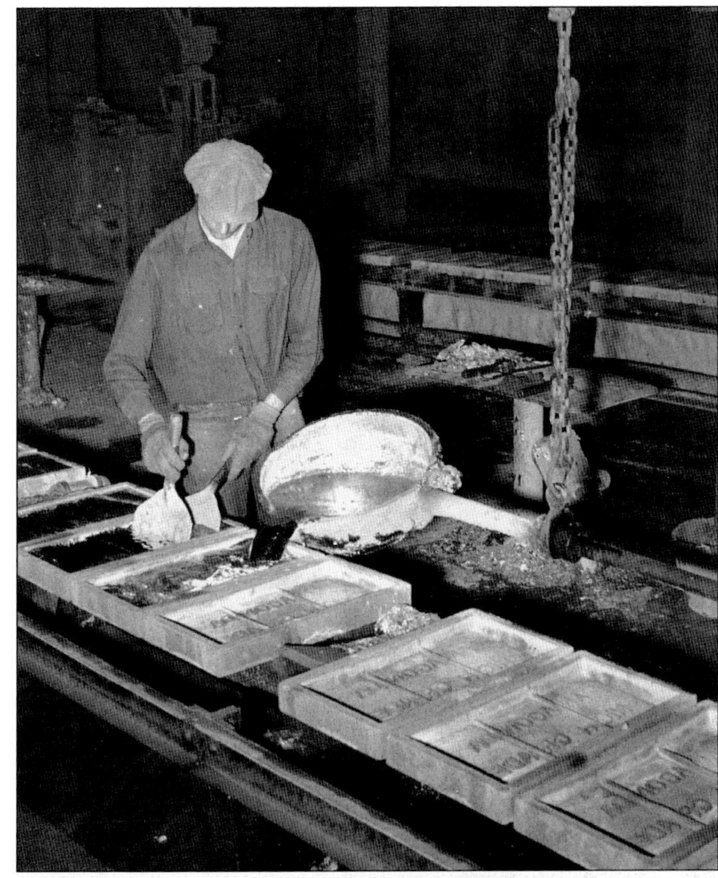

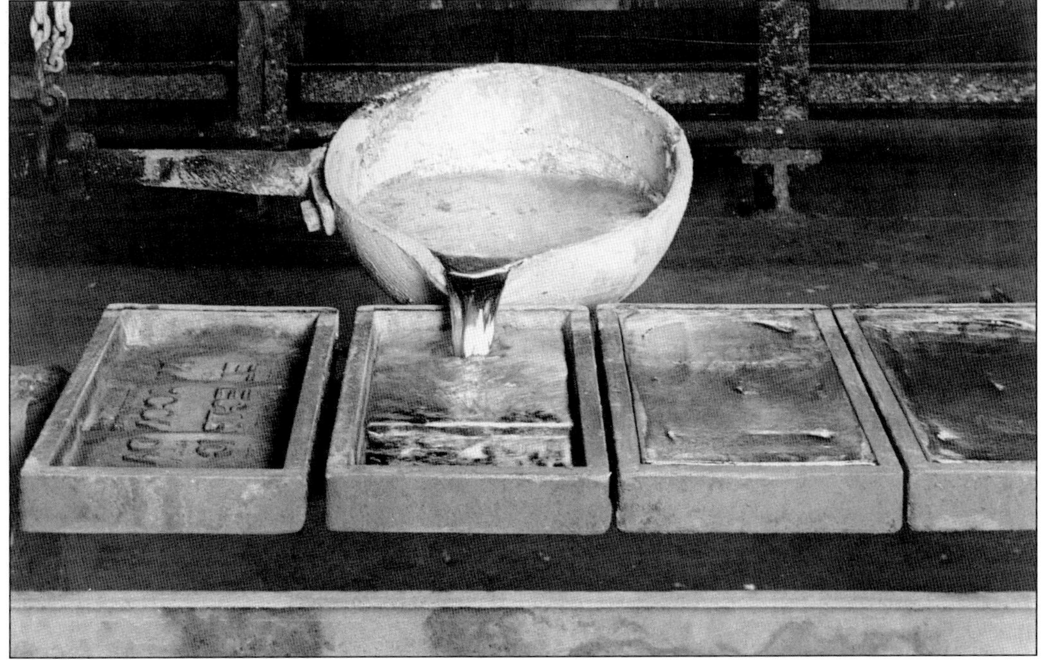

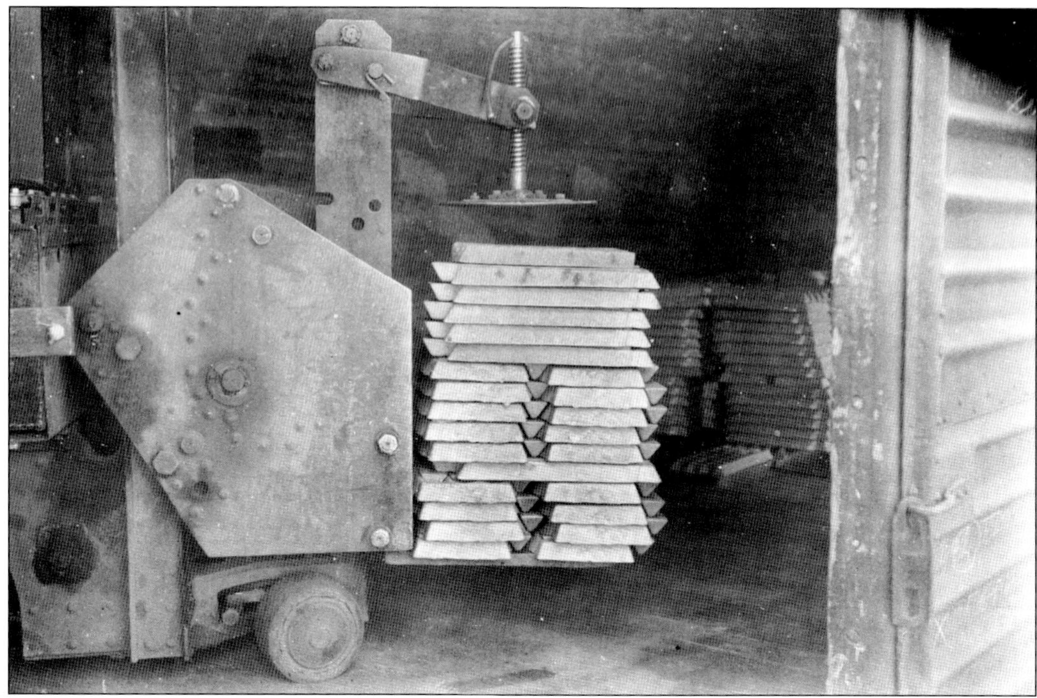

The zinc slabs, once cooled, were weighed either one or two at a time and had to be in a defined range of specified weight. The slabs were stacked in piles of 40 or 44, as seen above. They were then banded with galvanized steel and loaded into railroad cars for shipping, like in the image below. (Above, 1982.017.0066; below, 1982.017.0112a.)

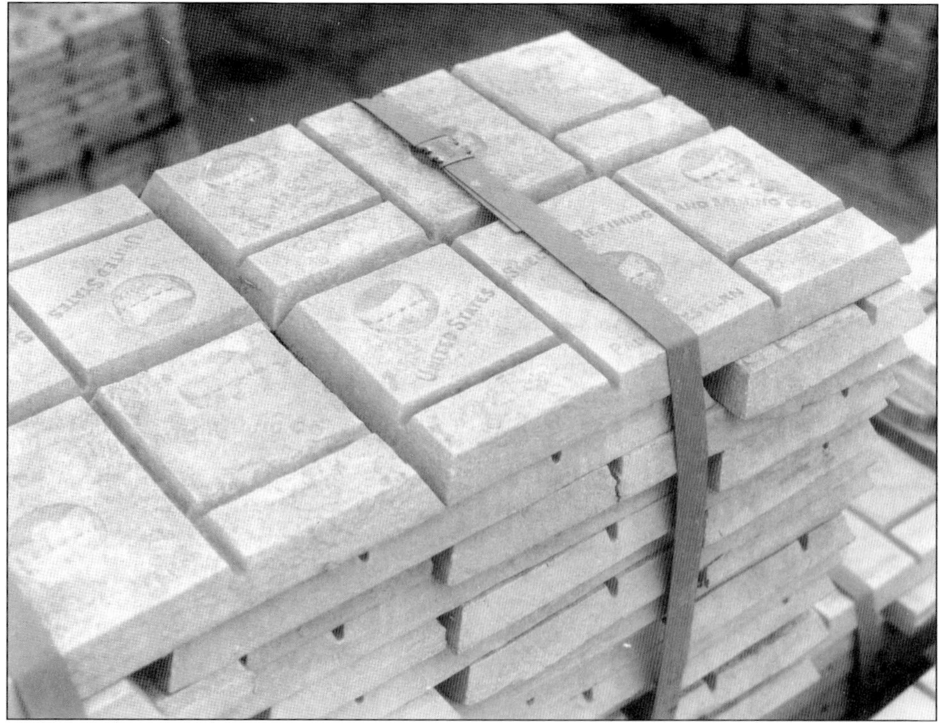

Four

THE ROD AND WIRE MILL AND OTHER DEPARTMENTS

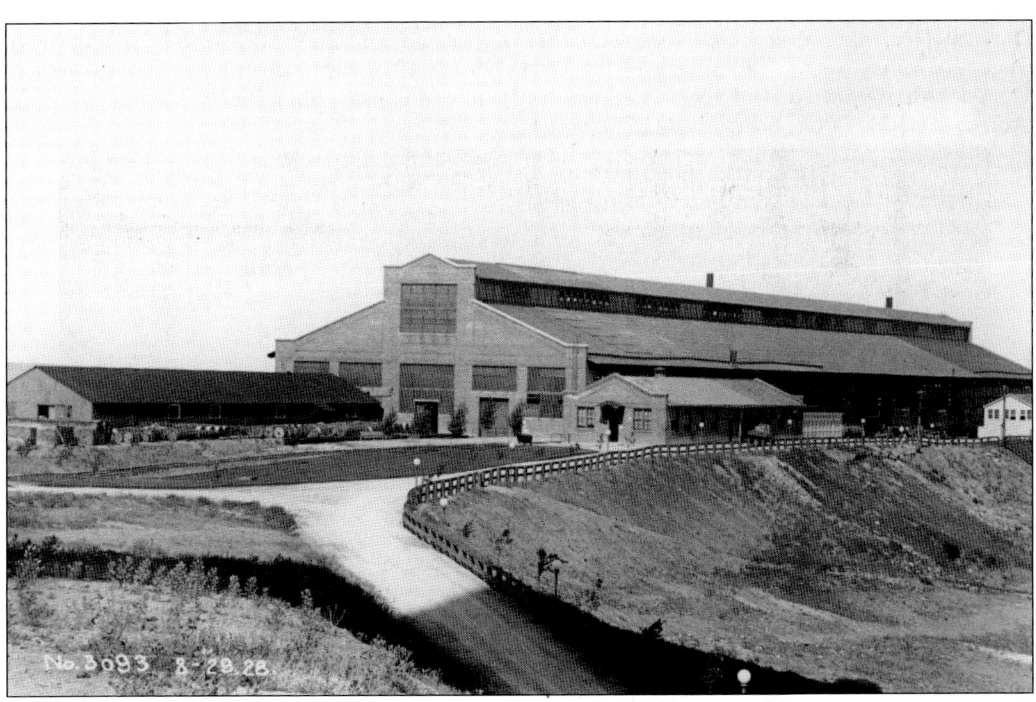

The Anaconda Copper Mining Company wanted to complete the circuit "from mine to consumer," so it created the Anaconda Wire & Cable Co. In May 1918, a copper rod, wire, and cable mill was completed at the Great Falls Reduction Department. The building was 650 feet long by 125 feet wide, as seen above. It was the first in the state and was prompted specifically by the Chicago, Milwaukee, St. Paul & Pacific Railroad, which wanted to electrify 656 miles of its line, and it was the first order the mill received. (From 2015.089.1.)

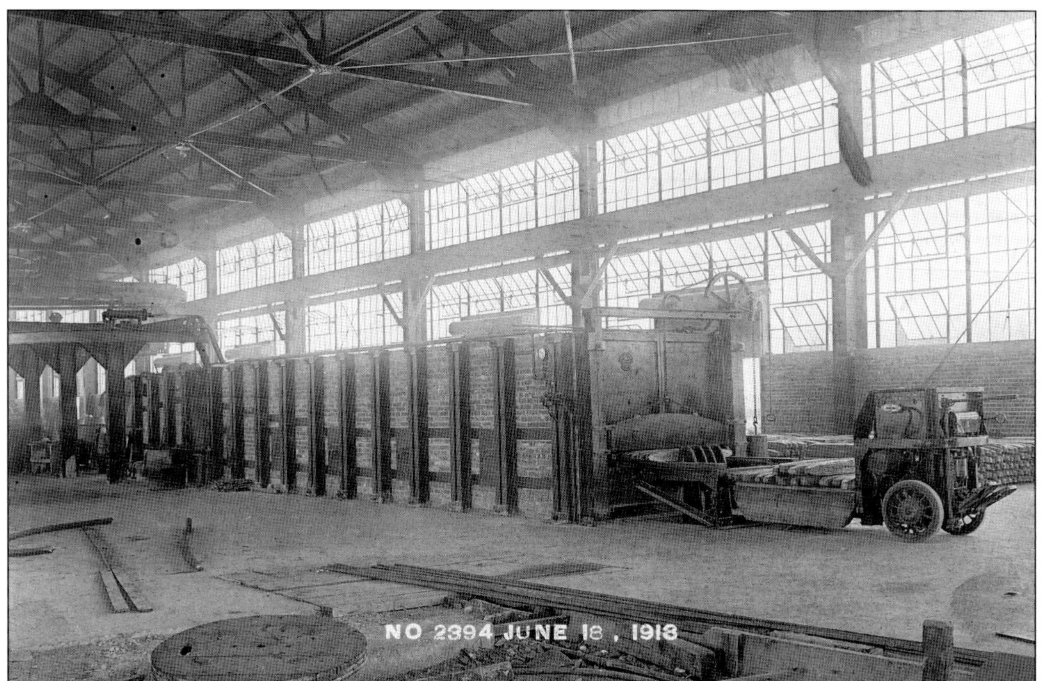

When the copper wire bars were transferred from the on-site copper refinery, they were placed at the start of the bar-heating furnace, as seen above and below. An electrical pusher moved the bars through the furnace at a rate of 150 bars an hour. When the bar reached the discharge door at the other end of the furnace, it was moved with a set of tongs to the roughing mill. The bar passed back and forth through the roughing mill, growing in length as it went. From the roughing mill, it went to the intermediate mill, where it continued to lengthen. (Above, 1982.017.0078; below, 1982.017.0080.)

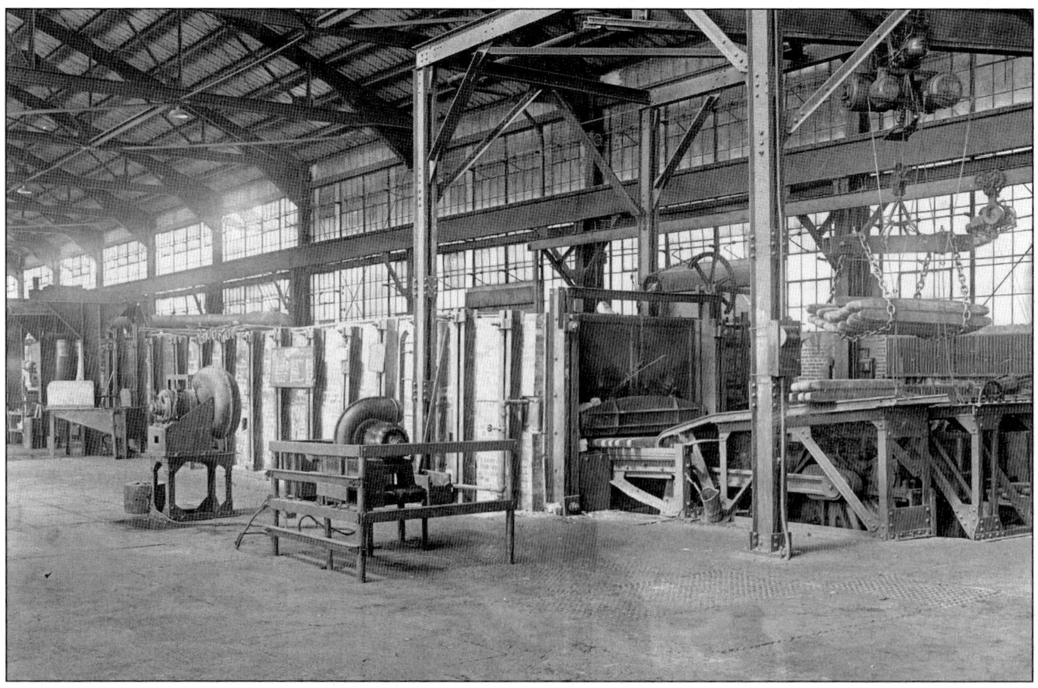

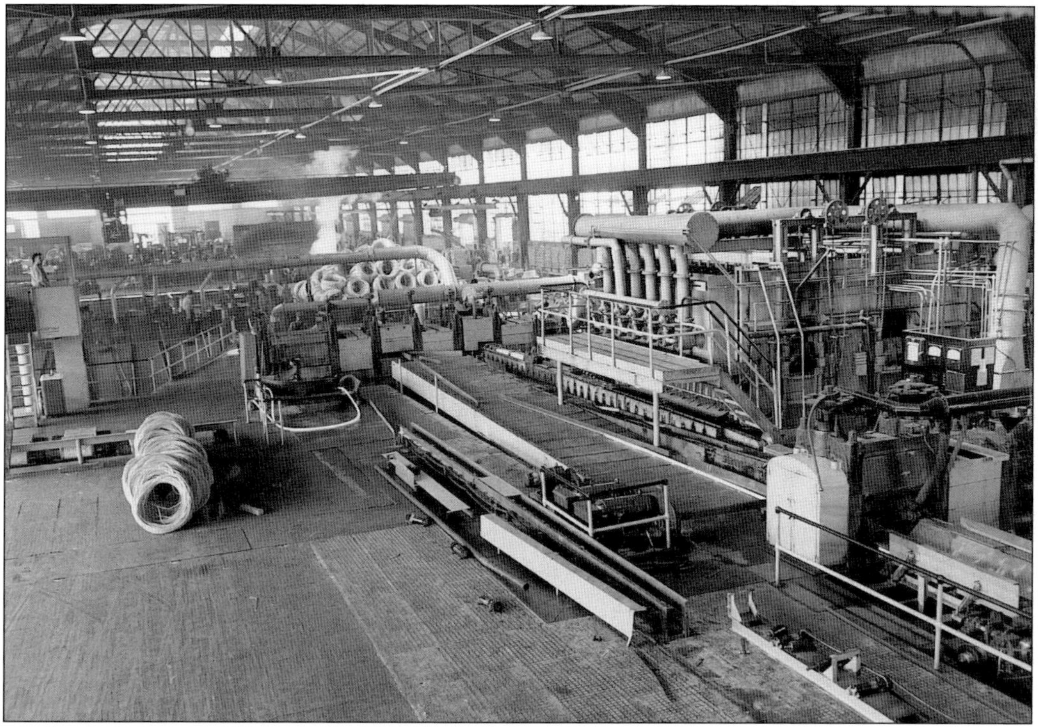

From the intermediate mill, the copper was fed into the finishing mill. The finishing mill made the rod its final standardized size, then sent it into the coiling machine. Within the Rod and Wire Mill Building, seen above and below, there were two rod mills, wire-drawing machinery, and stranding equipment. Rod Mill No. One produced rods from one-fourth to three-eighths of an inch in diameter. Rod Mill No. Two produced rods from three-eighths to one and a half inches in diameter, figure-eight rods, and flat rods up to three inches wide. (Above, 2023.043.0254; below, 1990.026.0275.)

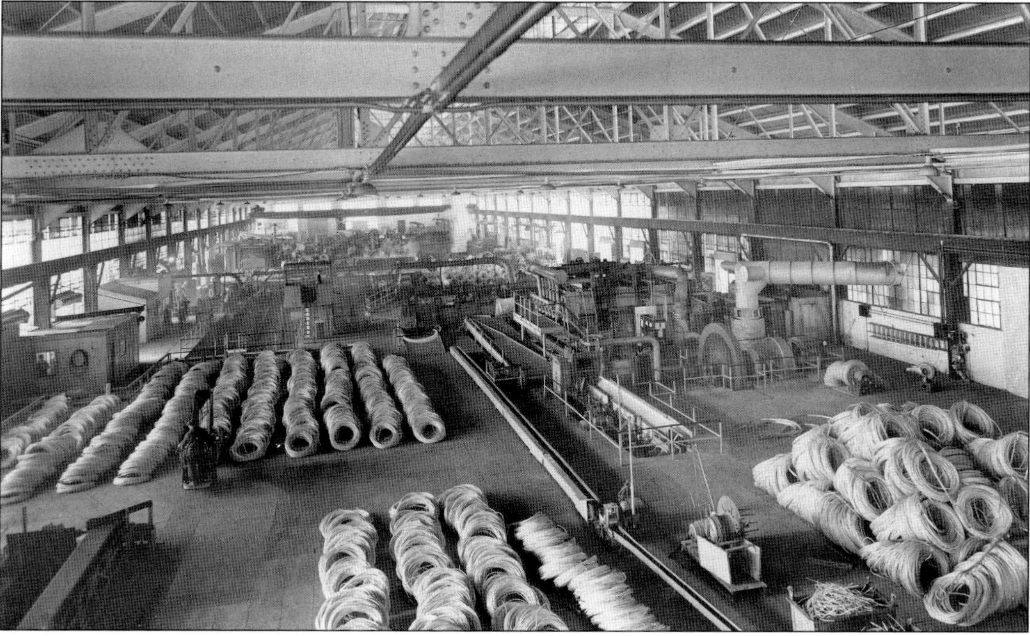

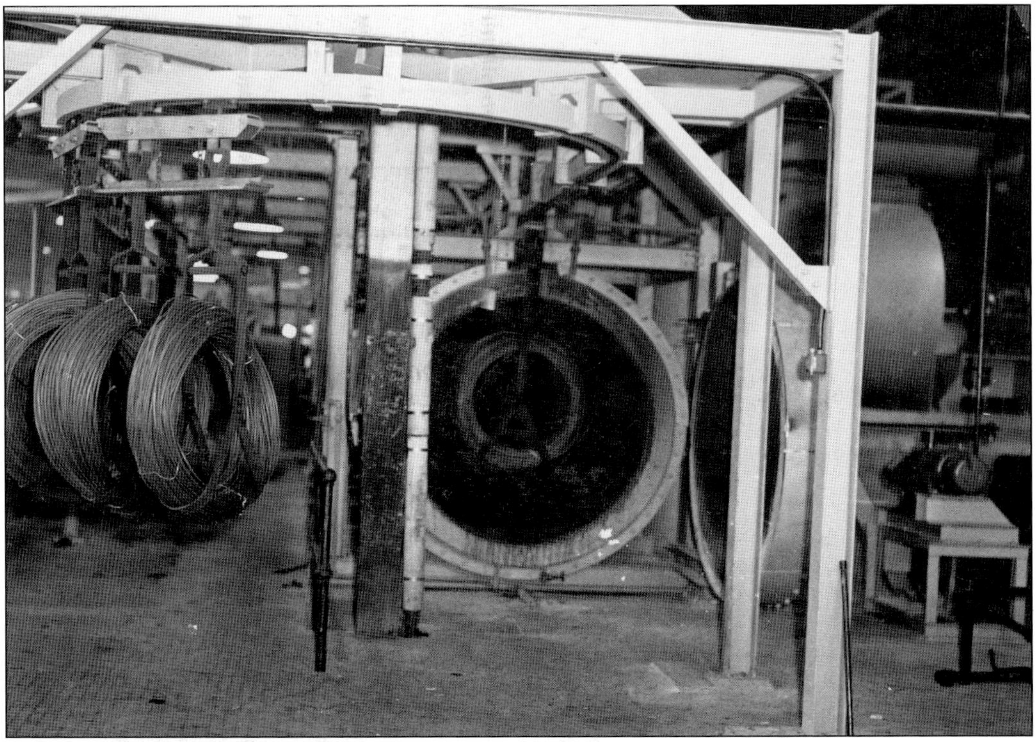

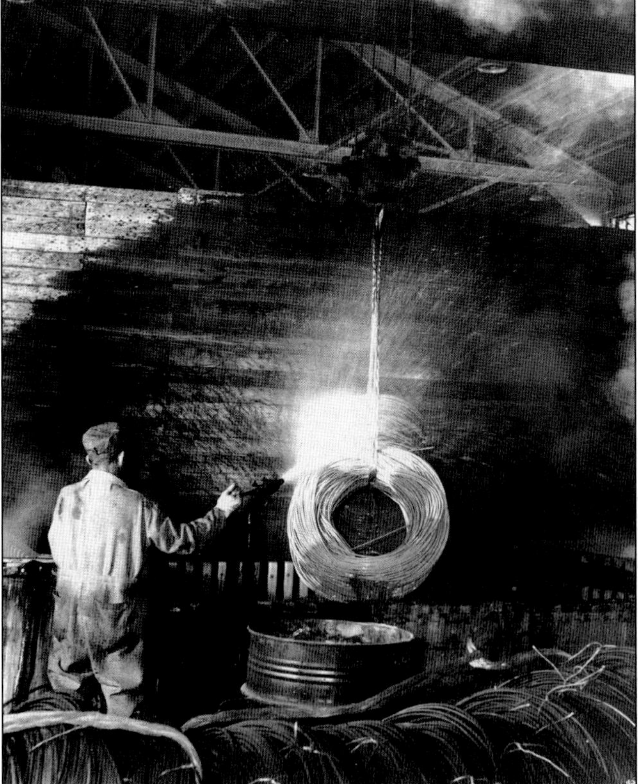

The finished rods came out black from oxidation in the milling machines. The oxide had to be removed before the rods could be drawn into wire. Once they went through the annealing machine, seen above, to make the copper more pliable, it was then pickled in a solution of diluted sulfuric acid. Once pickled, it was sprayed off with water, like Pat Boland is doing in the photograph at left. (Both, from 2015.089.1.)

The pickling took between 30 and 45 minutes, and the solution was heated to about 150 degrees. After washing with water, it was dipped in a boiling bath of soap and cream of tartar for 8 to 10 seconds, steaming in the image at right. This took care of any remaining acid and left a thin layer of soap to protect the copper from tarnishing. The rod coil was then put on a turntable for storage, as shown below. (Both, from 2015.089.1.)

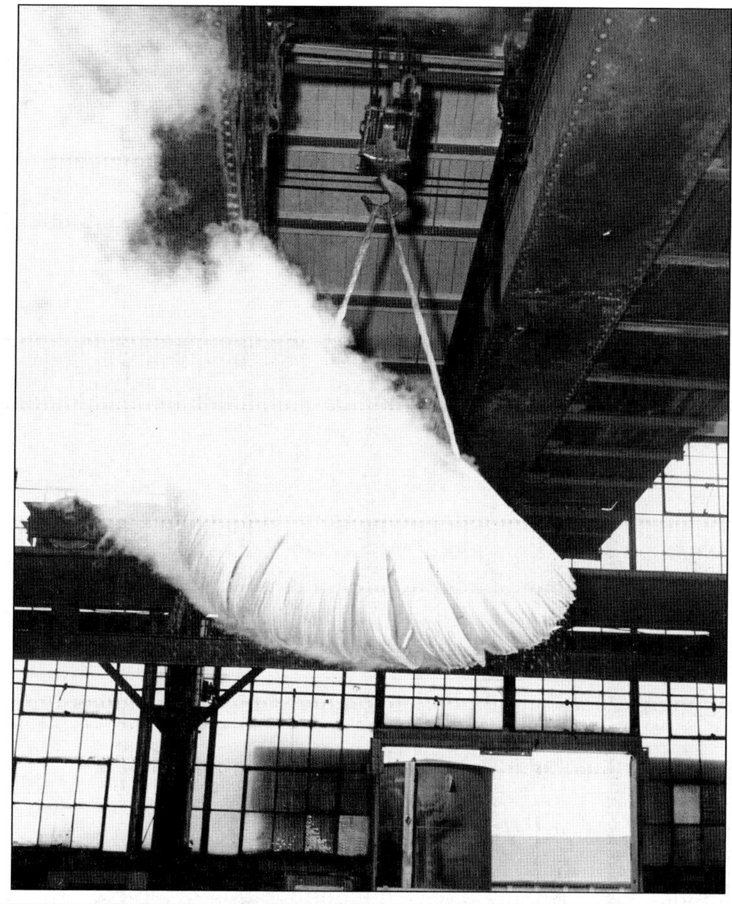

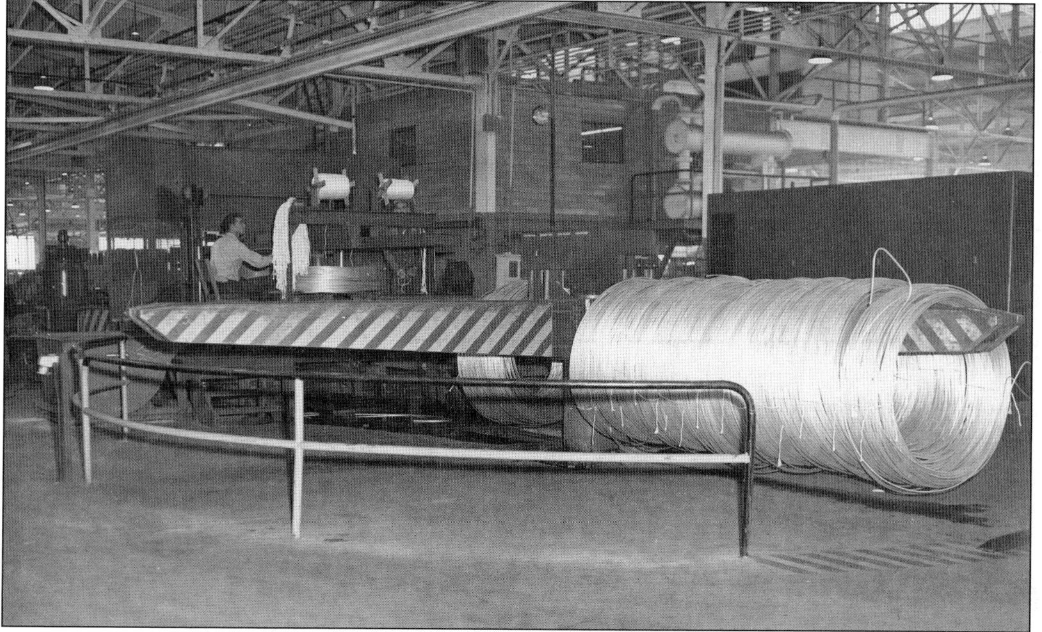

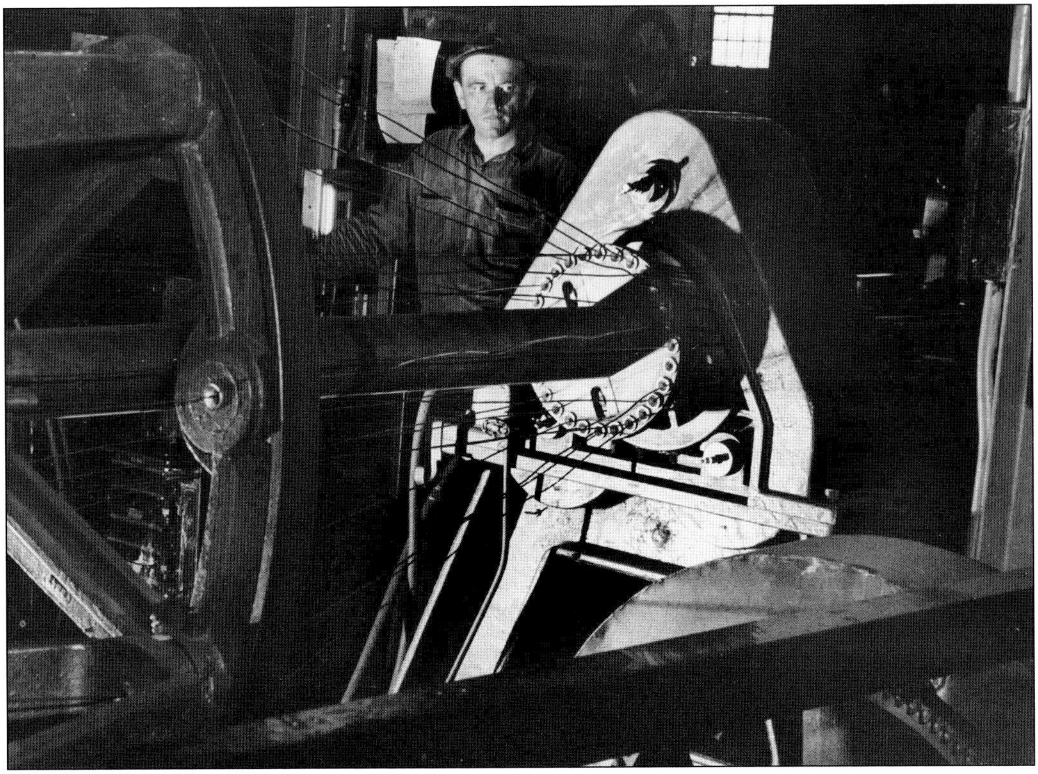

The rods were then sent to the wire-drawing machines. Once they were drawn into wire, the wire was wound onto reels, and they were placed in the stranding machines. Stranding machines can be seen above and below. Strand used for power transmission lines was made on a high-speed machine, which laid up six wires around a center wire. Depending on what size of wire or cable was needed, it could strand a cable of up to 127 wires. The revolving frames were driven through reversing and interchangeable gears. (Both, from 2015.089.1.)

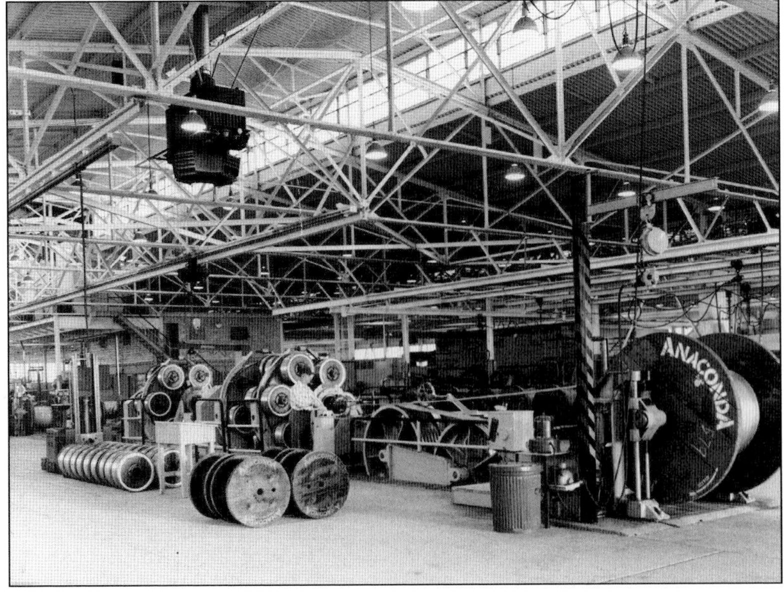

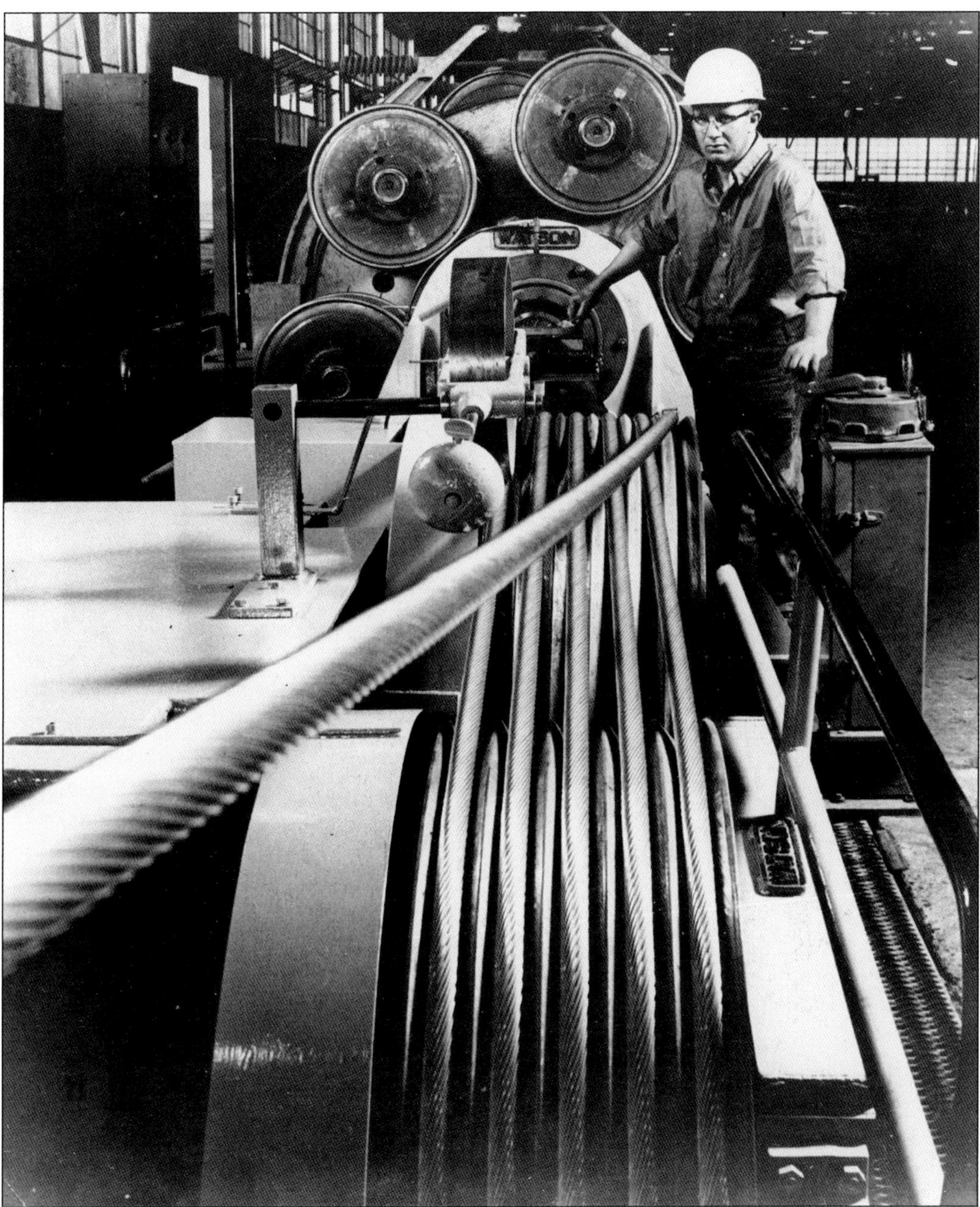

The Rod and Wire Mill produced three grades of hardness: hard, soft, and medium hard. The largest portion of the wire produced in either coils or bobbins for stranding was hard wire. Soft wire had the highest electrical conductivity. Most of the wire produced was used to make ordinary cable and hollow conductor cable, which was a special type of transmission cable. The hollow conductor cable was lighter in proportion to outside diameter than the ordinary type of cable made up of wires stranded about a spirally twisted copper core. Above, a man works at a cable-making machine. (1990.026.0557.)

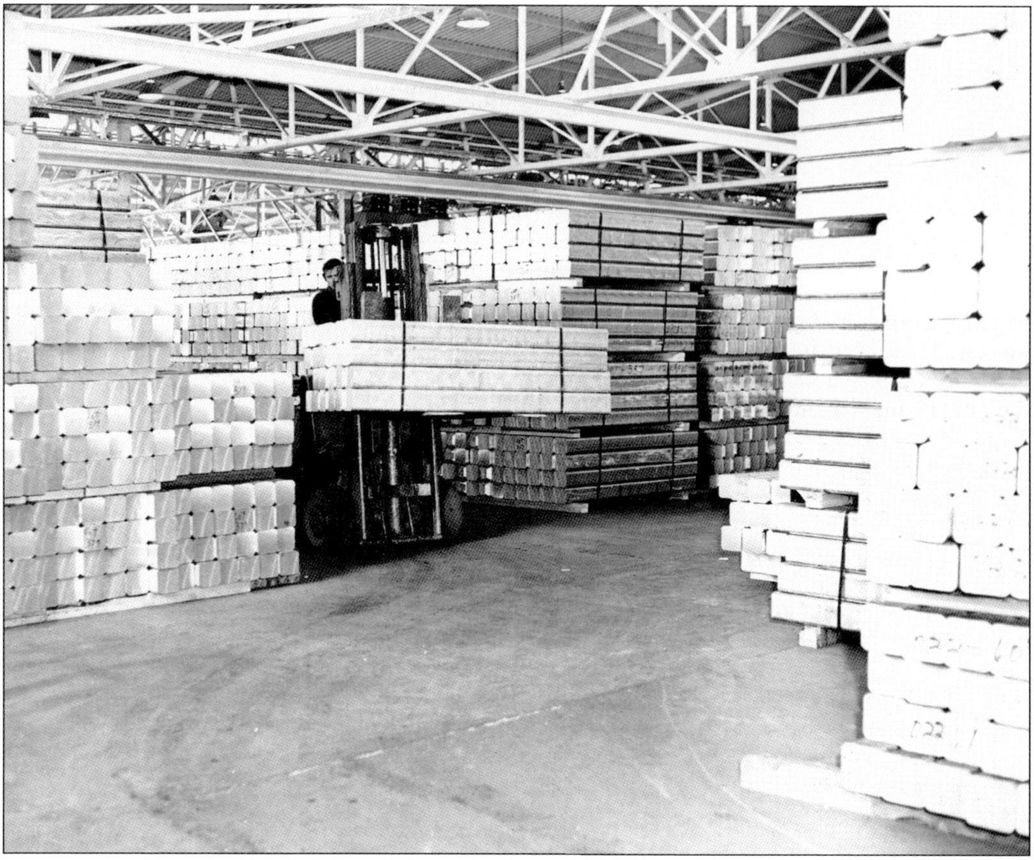

In 1954, the Anaconda Company began expanding and modernizing once again. It added aluminum processing by building a mill in Columbia Falls, Montana. With that came the addition of an aluminum rod mill to the Great Falls Plant, as seen below. Delivered all the way from Sweden, the mill arrived by boat then train on August 25. The million-dollar machine took 15 freight cars and was compromised of 120 separate crates, some weighing as much as 10 tons. When it arrived, it was put together under the supervision of Swedish engineers. Above, aluminum bars arrive from Columbia Falls. (Both, from 2015.089.1.)

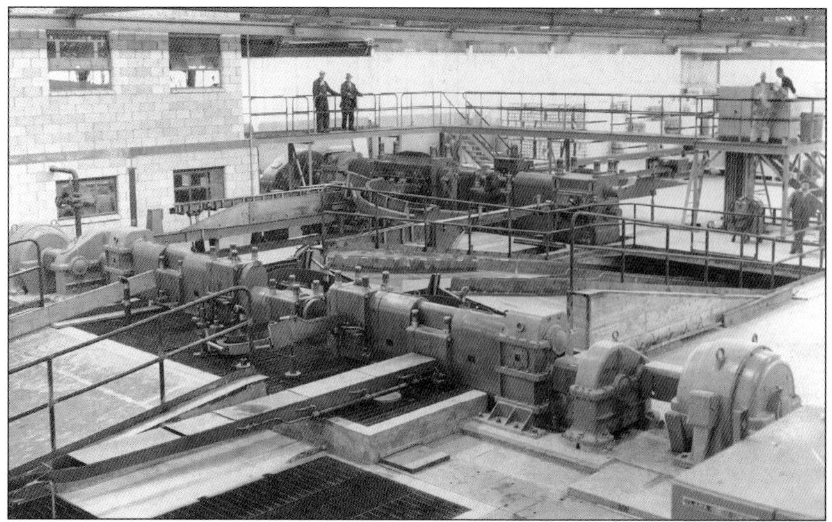

In 1923, a separate department was established for the processing of cadmium that came out of the Zinc Plant. At right, Carl Trip collects the residue cakes from the Zinc Plant's leaching division and takes them to be dried in roasters. Fifty percent of that residue is cadmium. The cadmium was then purified and electrolyzed in cells to produce sheets. Those sheets were taken over to the Cadmium Plant, seen below, and melted down and cast in various shapes for market. (Both, from 2015.089.1.)

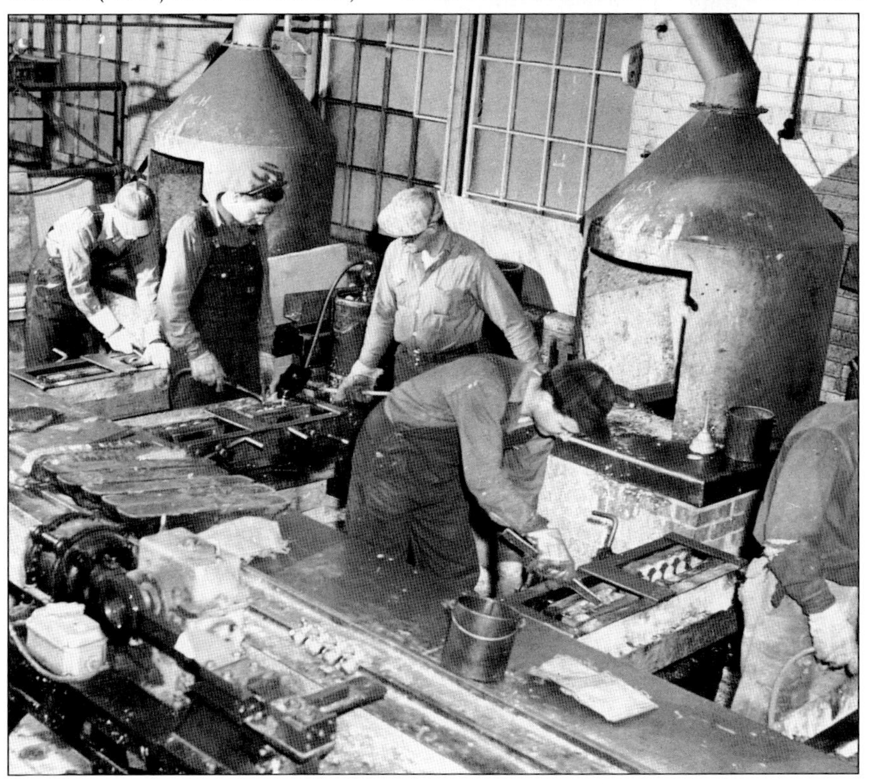

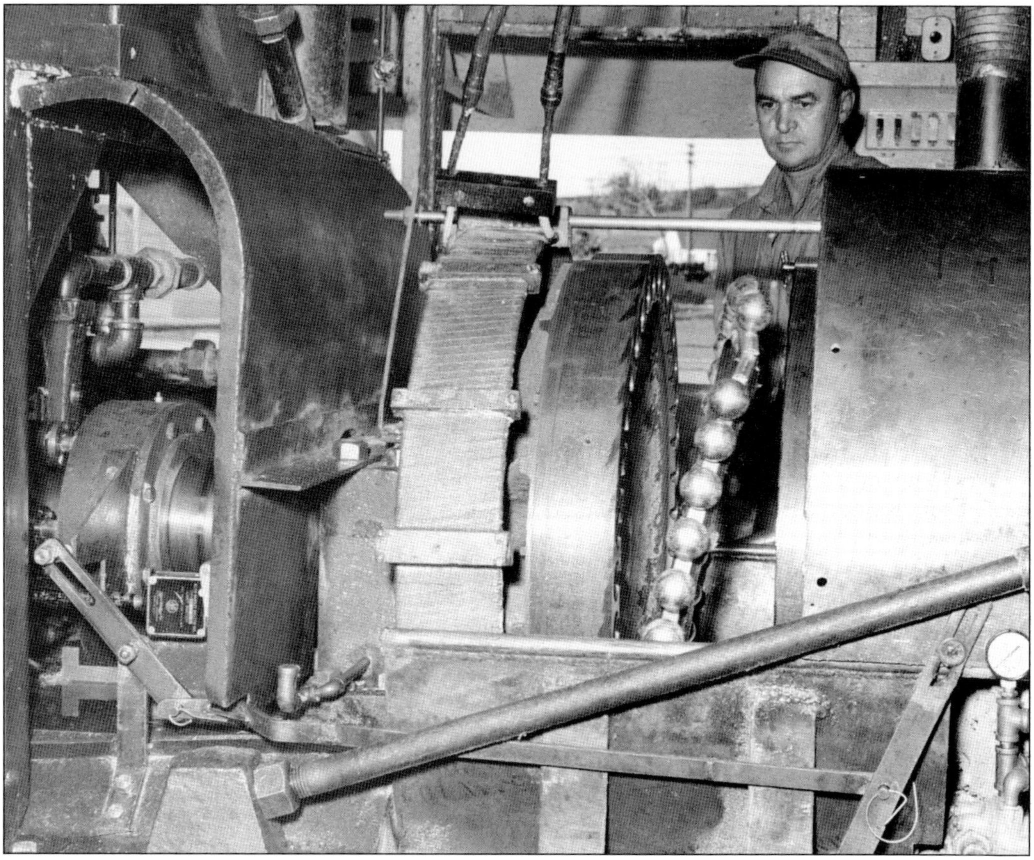

Cadmium is used chiefly in the manufacture of bearings, electroplating, and the production of yellow pigments. It was cast in balls, slabs, and sticks for market. Above, Glen Sigler works the cadmium centrifugal ball casting machine, and below shows the results. Approximately three to five tons of cadmium were produced a day. Indium was also recovered from the zinc residue and is used for protecting bearings from oil corrosion and the production of non-tarnishing silver. (Both, from 2015.089.1.)

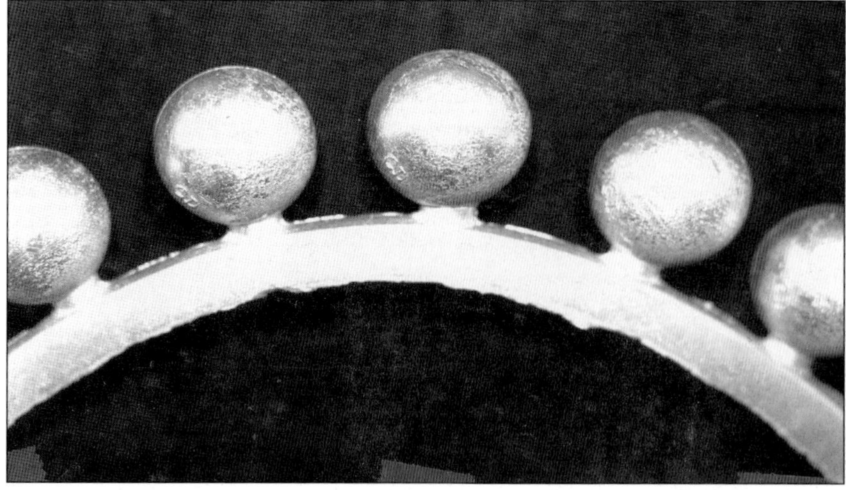

The unsung hero with which the smelter could not have been so innovative, let alone flourished, was the Electrical Department. In the earliest years of the Great Falls Plant, all electrical power was developed from the Black Eagle Falls and Dam. As the years went on, power was generated by the Montana Power Company using Black Eagle Dam, Rainbow Dam, Ryan Dam, and Morony Dam. The energy was distributed from three substations: the Zinc Plant substation, Copper Plant substation, and Rolling Mill substation. At right is the Black Eagle pump house substation, and below is the Rainbow powerhouse just after it was built in 1910. (Right, 2022.028.0019; below, 1984.007.0001m.)

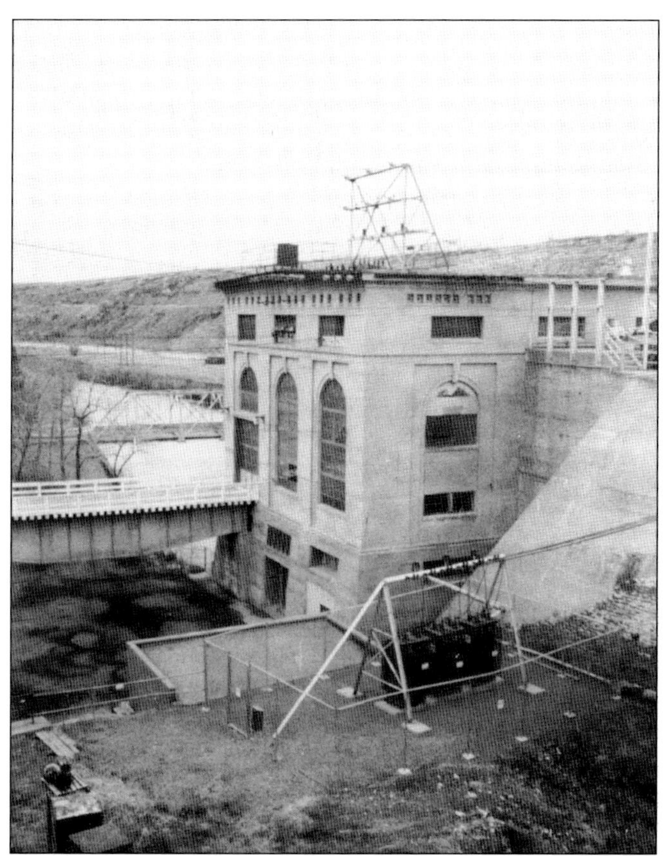

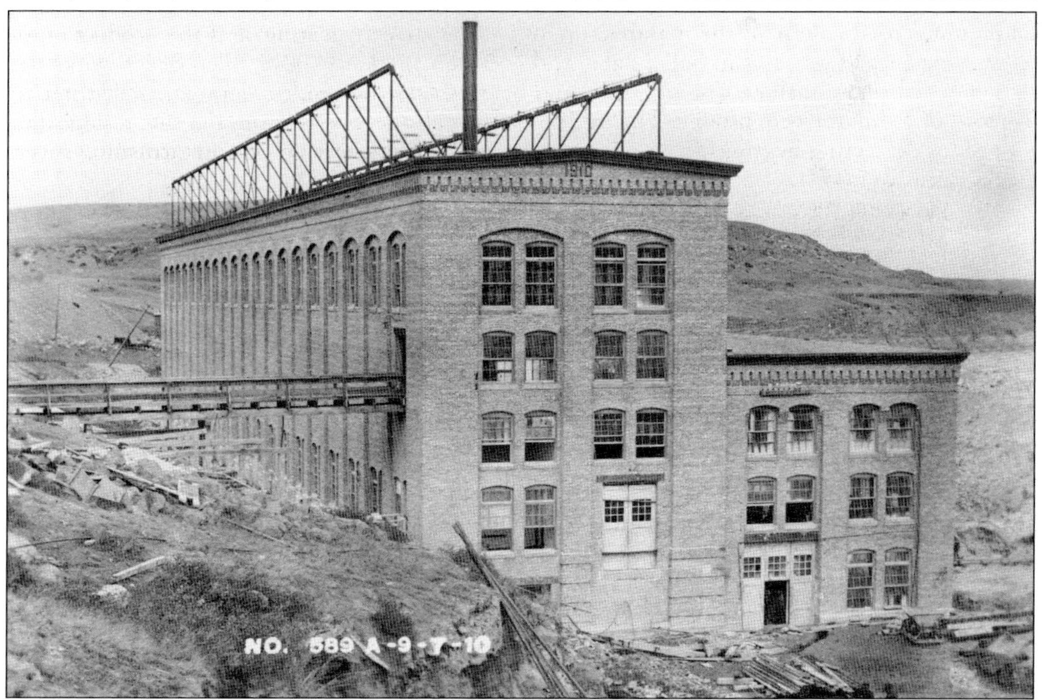

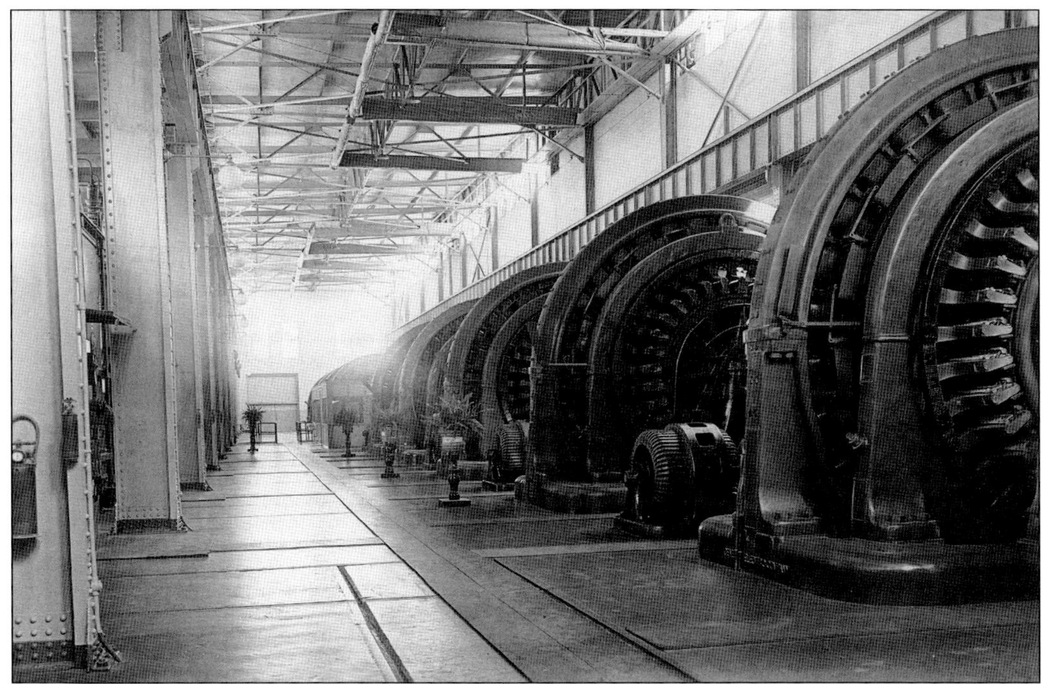

Inside the Zinc Plant substation there were eight mercury-arc rectifiers, nine transformers, and nine rotary converters, which can be seen in the above image by Yaw Photo Studio. The switchboard room, seen below, was at the center of the station, where it was well ventilated with humidified air. The substation supplied power directly to the zinc electrolytic cell room, the Cadmium Plant, and other motors in the plant. The total capacity of the Zinc Plant substation was 73,000 kilo-volt-amperes or 98,500 horsepower. (Above, 2024.047.0007; below, from 2015.089.1.)

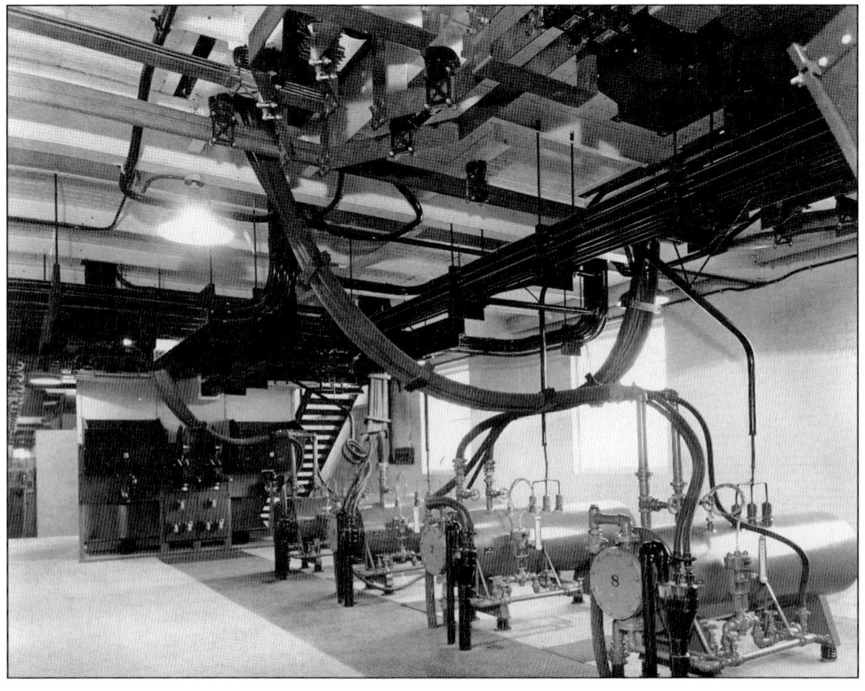

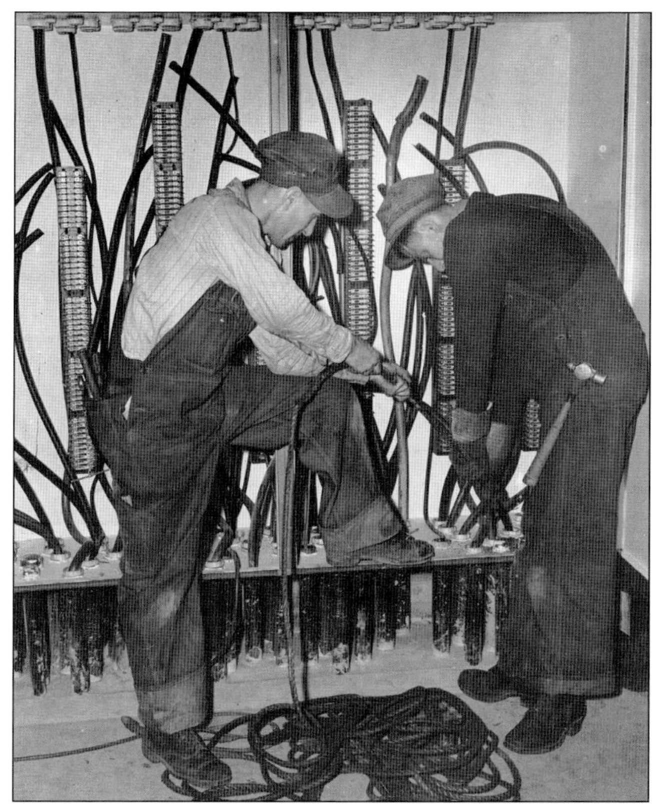

The Copper Plant substation's capacity was 13,000 kilowatts or 17,450 horsepower. The capacity for the Rolling Mill substation was 4,884 horsepower. The mercury-arc rectifiers were added to the Zinc Plant substation during World War II to help meet the high demand. At right, Rudy Polich (left) and Ray Stanich (right) pull wire into a junction box. In 1956, germanium rectifiers were added to the Copper Plant substation on a separate circuit for the starting sheet section, as seen below. (Right, from 2015.089.2; below, from 2015.089.1.)

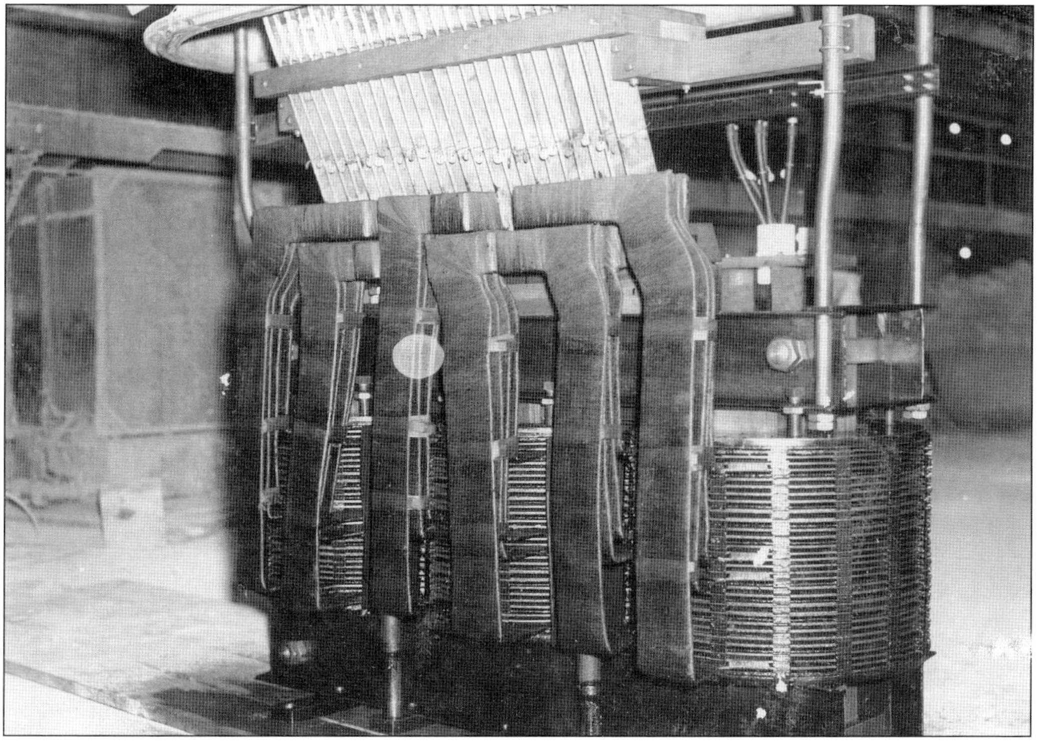

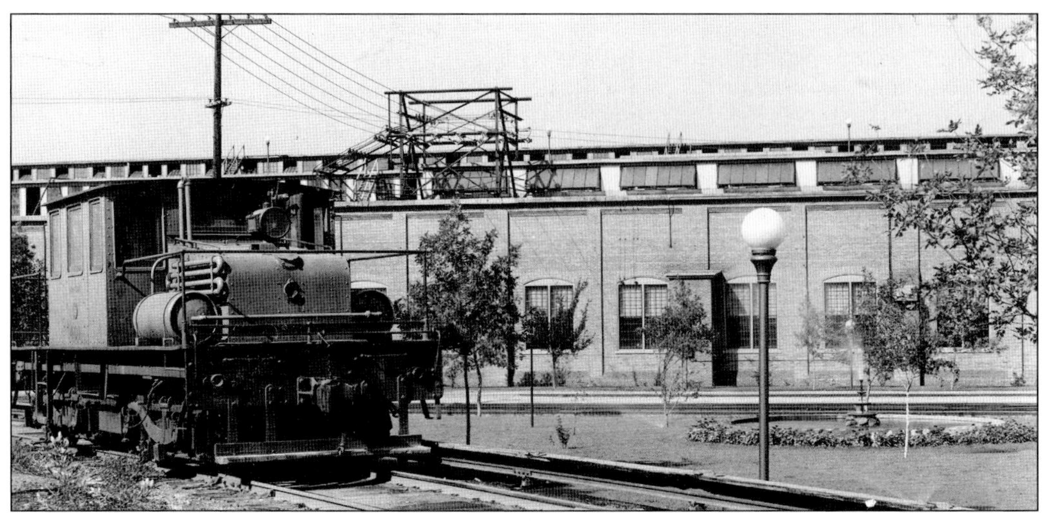

The electric tram system moved all the materials around the smelter, as seen above. Starting with the Great Northern Railroad tracks at the skyline and low line, the plant had a complete belt line with two transfer points for interplant switching and delivery of materials. There were nine different engines to move cars around, a couple of which are shown below. Nearly all the tonnage was hauled up grade, averaging 1,500 tons a day. (Both, from 2015.089.1.)

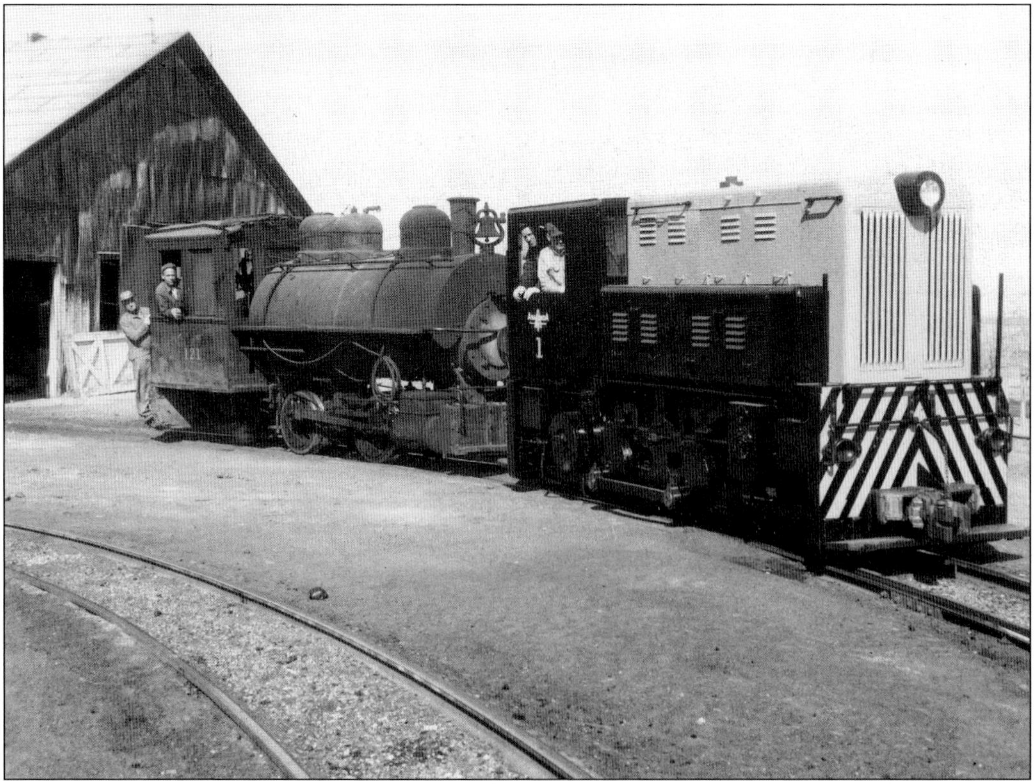

There was, of course, a laboratory at the plant, which analyzed materials as they came to the plant as well as doing control work for every step of the various plants' processes. A certificate of analysis was shipped along with the product to the consumers. Featured at right is an Arsine detector. (From 2015.089.1.)

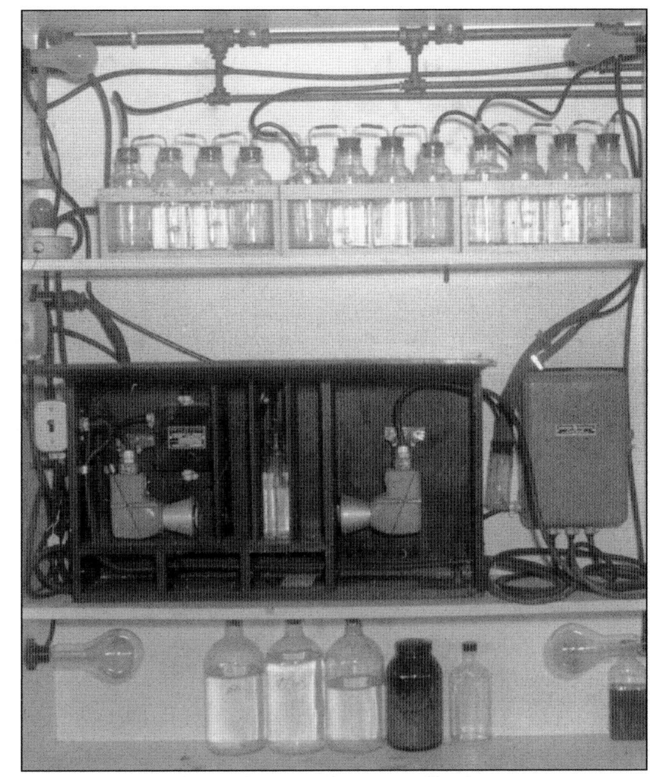

During World War I, the government requested that the Anaconda Company produce ferromanganese. Five electric furnaces were installed, and three were producing when the Armistice was signed. With the demand gone, the furnaces were shut down and dismantled. Below is one of the ferromanganese furnaces. (From 2015.089.1.)

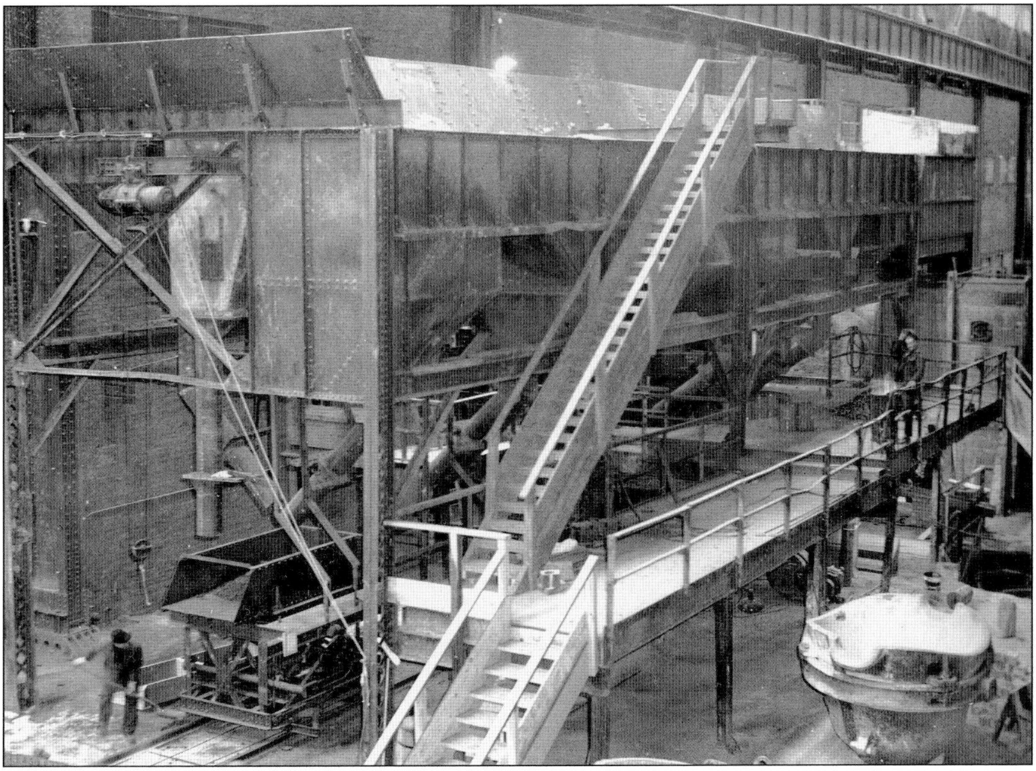

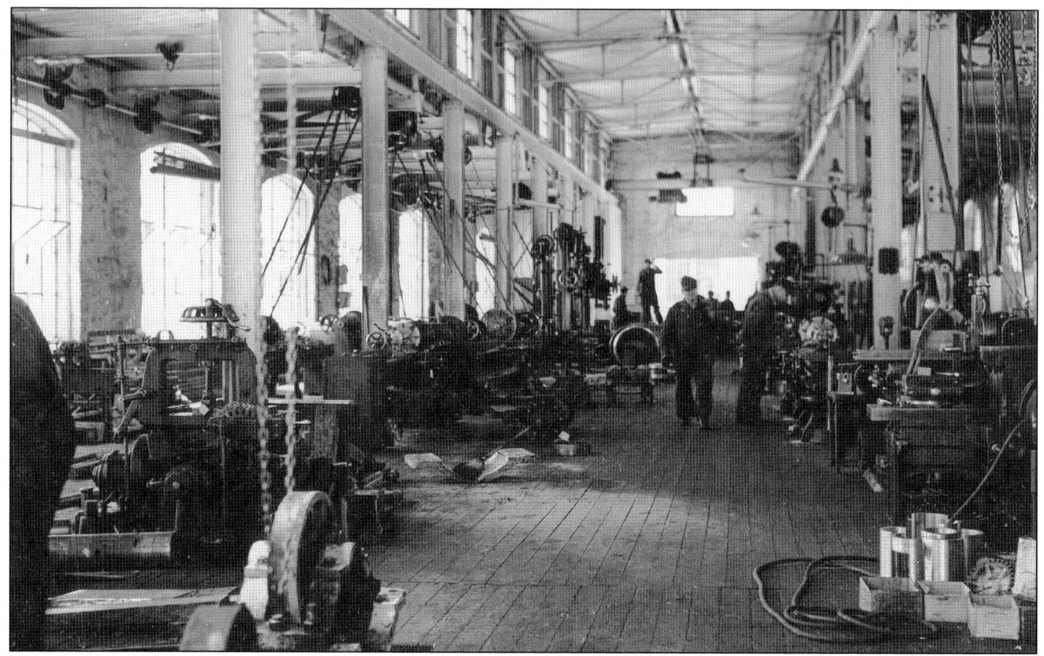

The Anaconda Company made sure that it was able to be self-sustaining by having various support shops and personnel. The machine shop, seen above, and boiler shop could produce anything to repair, build, or create anything needed for the operations to continue. Security watchmen like Bert Higgins, featured below, patrolled day and night to keep everyone safe. (Both, from 2015.089.2.)

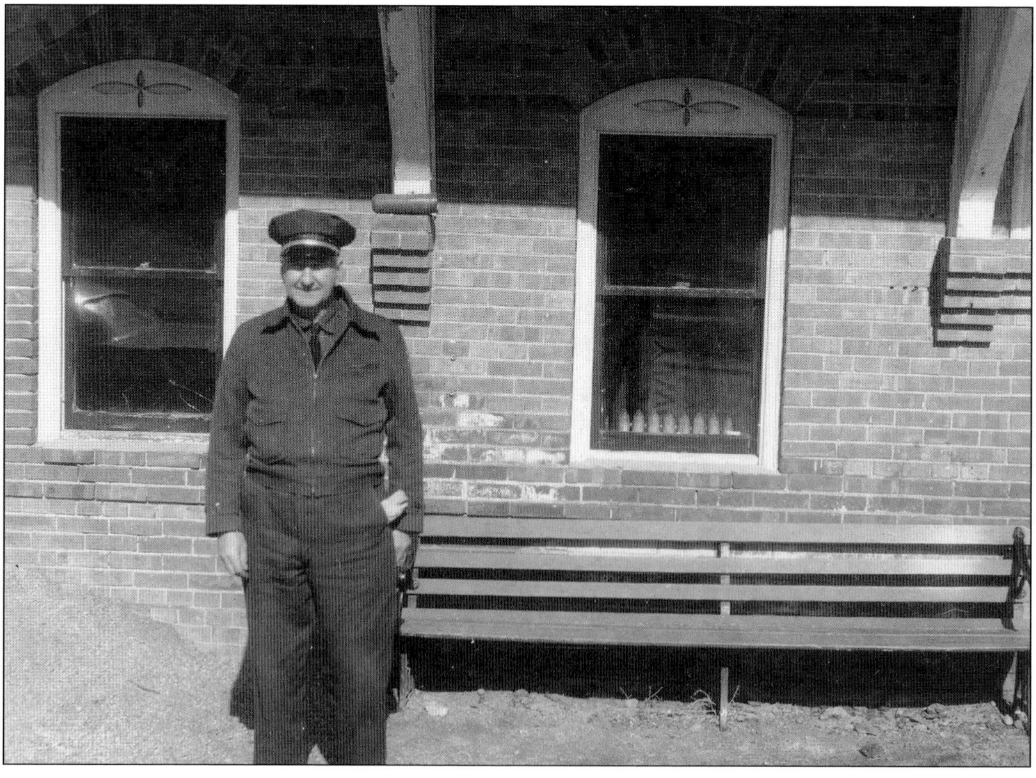

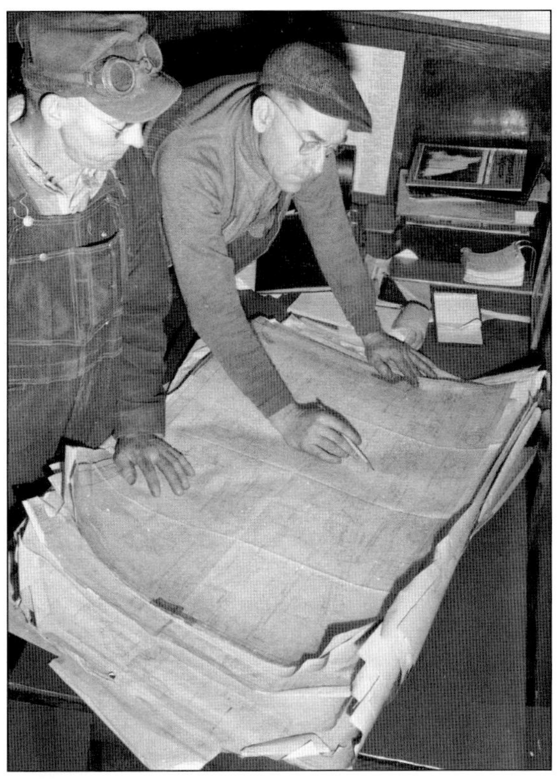

There were also shops for blacksmithing, foundry work, brick masonry, carpentry, painting, piping, and patterns. Each shop was equipped with the necessary tools and equipment for the work and a skilled corps of employees. Most shops were housed in separate buildings and were under independent supervision. At right, Jess Tapp (left), outside foreman, and Charles "Chick" Meyers (right), shop foreman, look over plans in the boiler shop. Below, Chick Meyers is shown working the cathode header in the boiler shop. (Both, from 2015.089.2.)

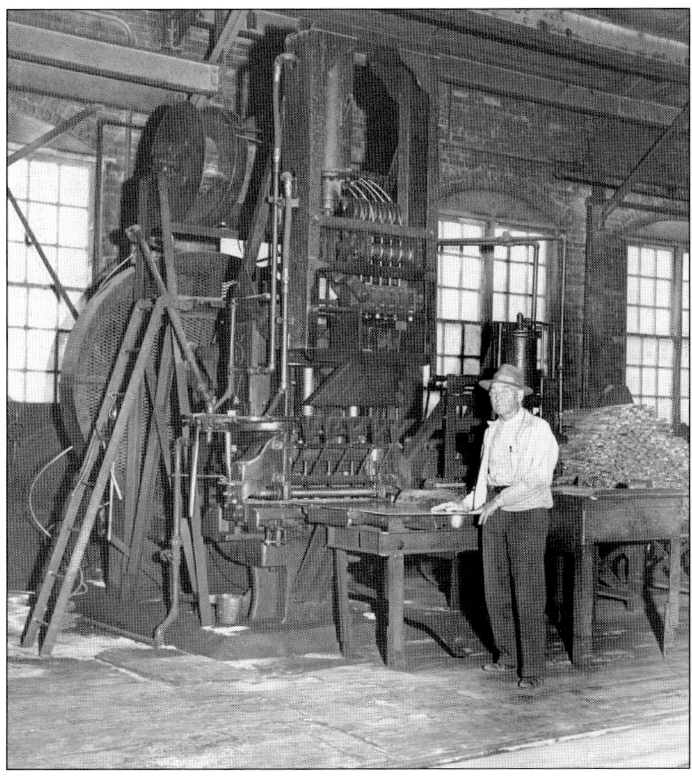

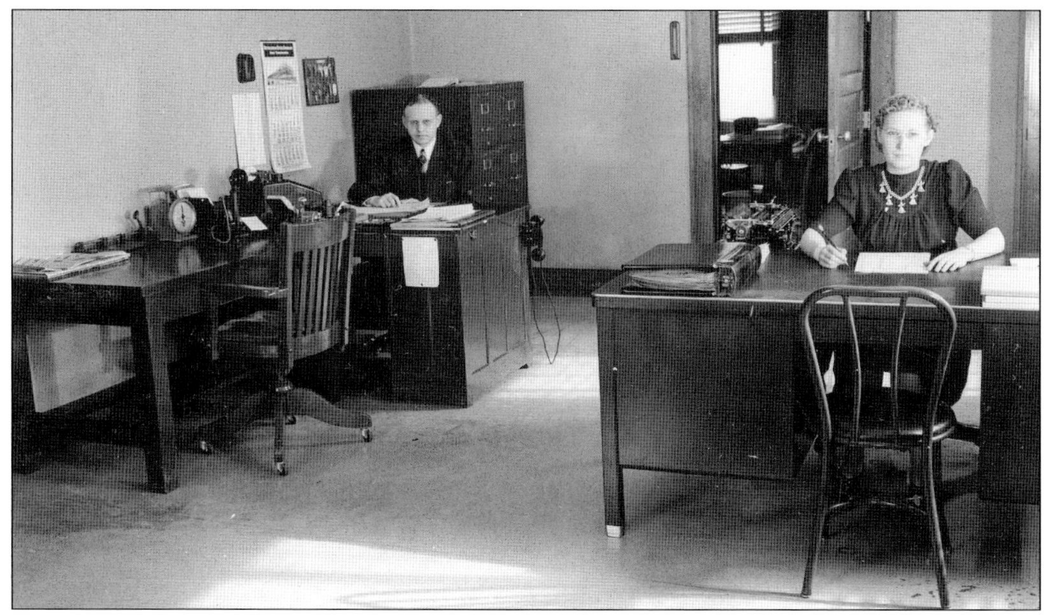

Of course, those behind the scenes cannot be forgotten, such as the paper pushers who kept the business end of the company going. They made sure everyone got paid, trained, taken care of, and scheduled. There was even a safety and welfare engineer who knew each employee personally to make sure everyone was doing well. Above are Robert McCollough and Marjorie Liggett in the data processing department, and below is Albert F. "Bud" French, who was a safety and welfare officer. (Both, from 2015.089.2.)

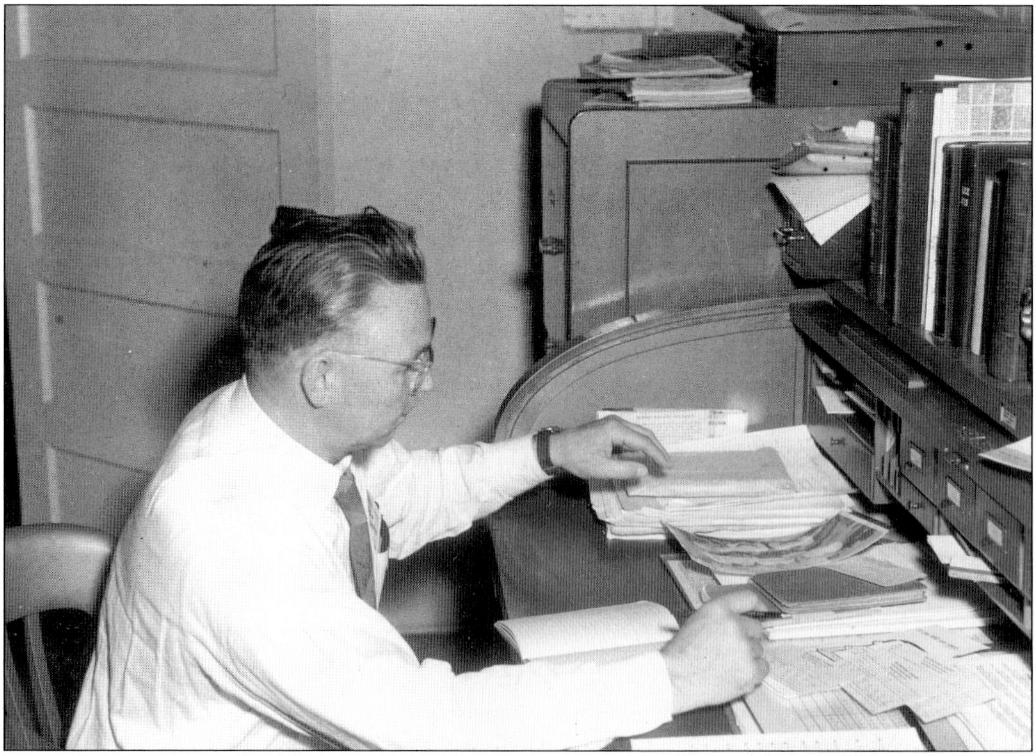

Five

Company Life and the People

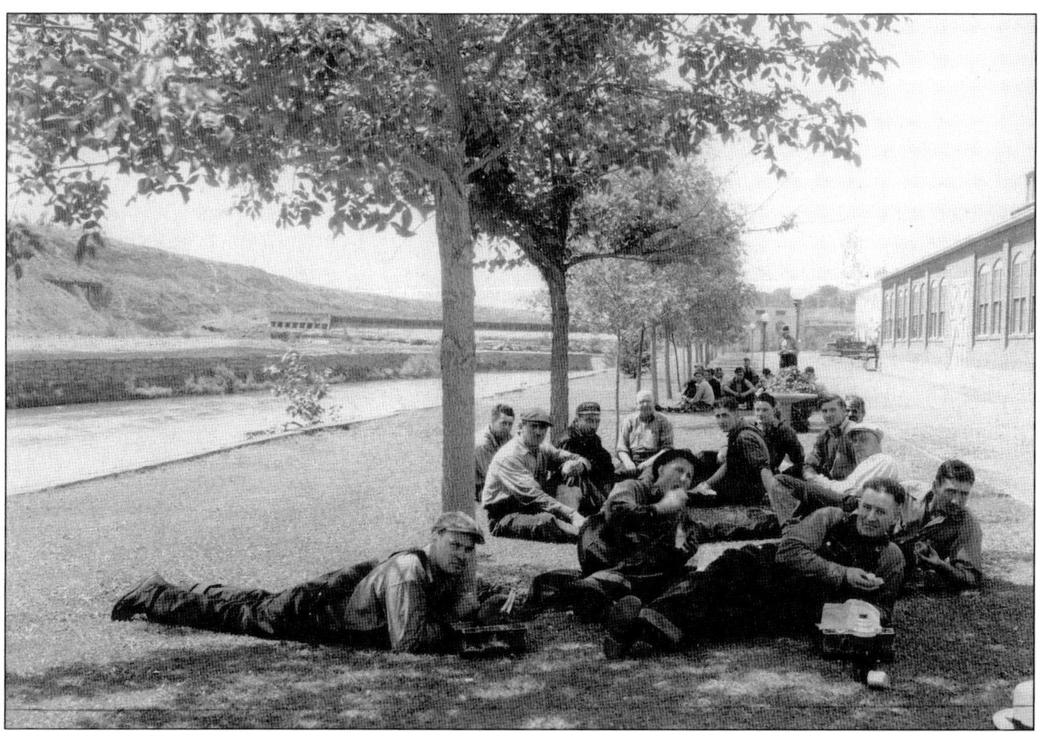

The Anaconda Copper Mining Company valued its people. It strove to provide for workers' families and their happiness. When Anaconda took over the Great Falls Reduction Department, the grounds of the plant were spiffed up to be pleasant for the workers and to appeal to visitors, especially along the Low Line Road. In fact, the plant became a tourist attraction for the area. Above, men from the boiler shop enjoy their lunch in the shade of some trees. (1982.017.0116.)

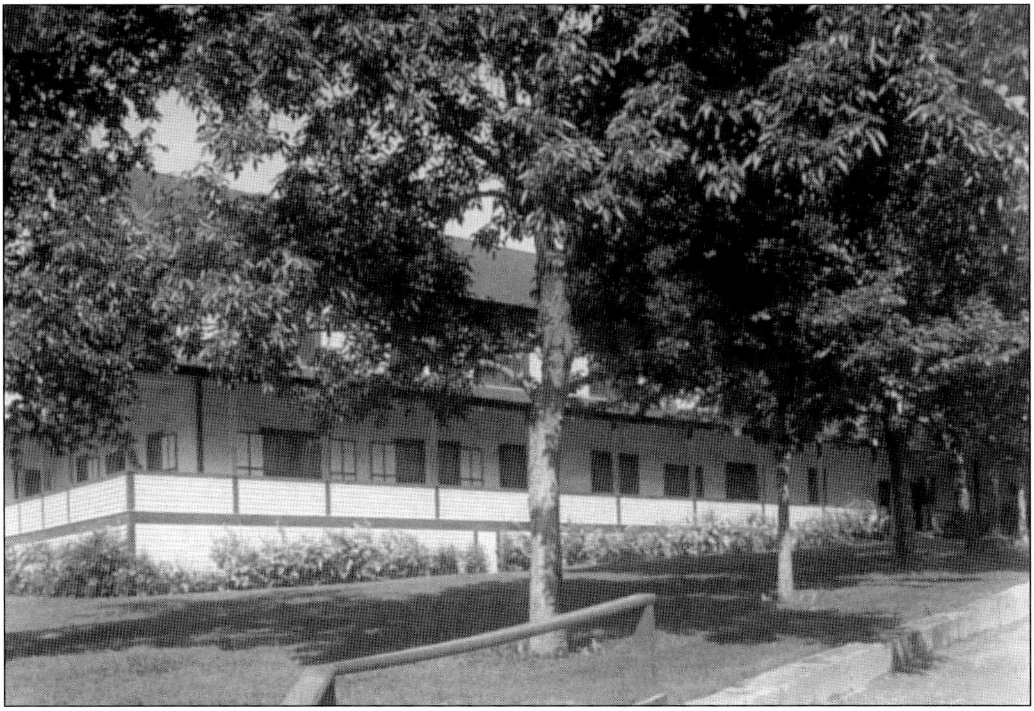

Depending on the year, the first thing seen upon reaching the company grounds was the employee club. The original clubhouse, seen above, had six bowling alleys, billiard tables, horseshoe courts, a rifle and revolver range, and a library. It was situated farther into the plant, near the supervising housing. Sadly, it was destroyed in a fire in 1943. The second clubhouse was built in 1950 and still stands as the Black Eagle Community Center, featured below. There was an open house when it opened, putting on full display the new luxury. There were eight bowling alleys, locker rooms, a lounge, a kitchen, and an assembly room. (Above, 2023.043.0180; below, from 2015.089.1.)

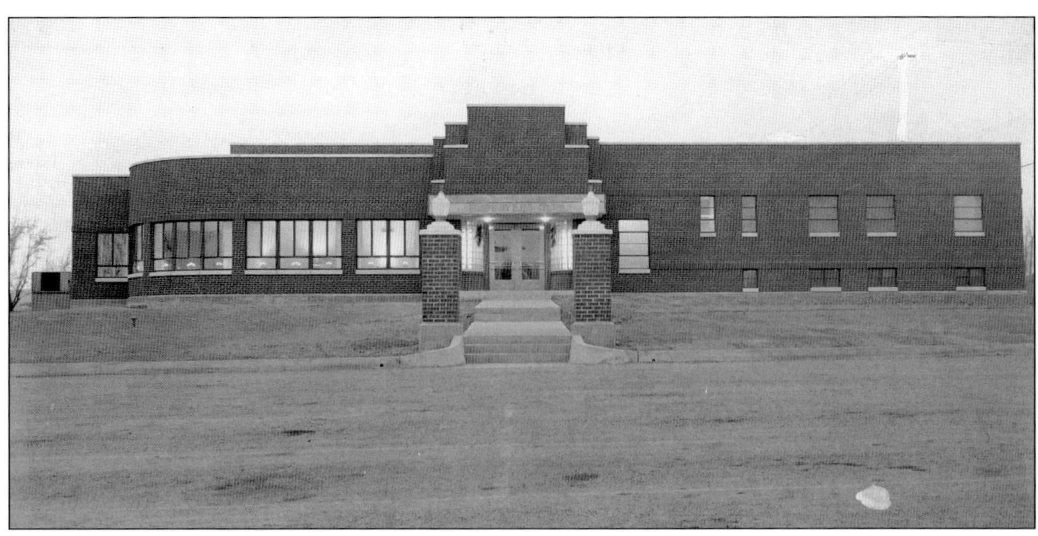

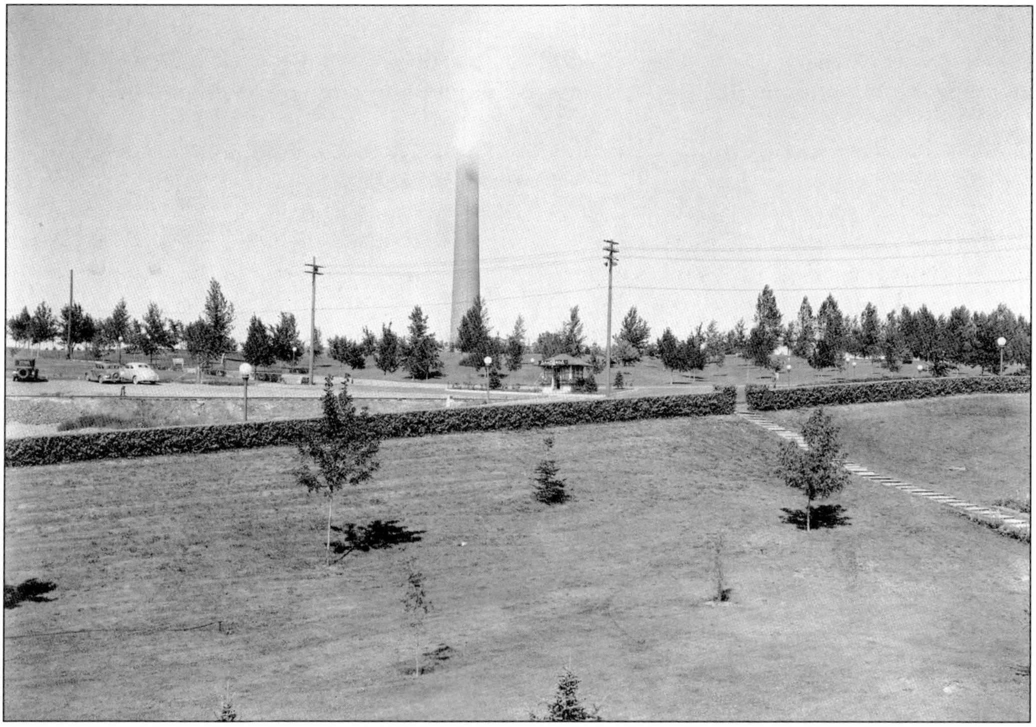

Also, before reaching the guard station, the Anaconda Company Golf Course is visible above. Established in the 1920s as a nine-hole golf course, it was near the brick department, as shown in the photograph below. It was beautifully maintained with sand greens. When the company closed, the golf course was gifted to the City of Great Falls, and it has since had grass greens and another nine holes added. (Above, 1982.017.0292; below, 1982.017.0090.)

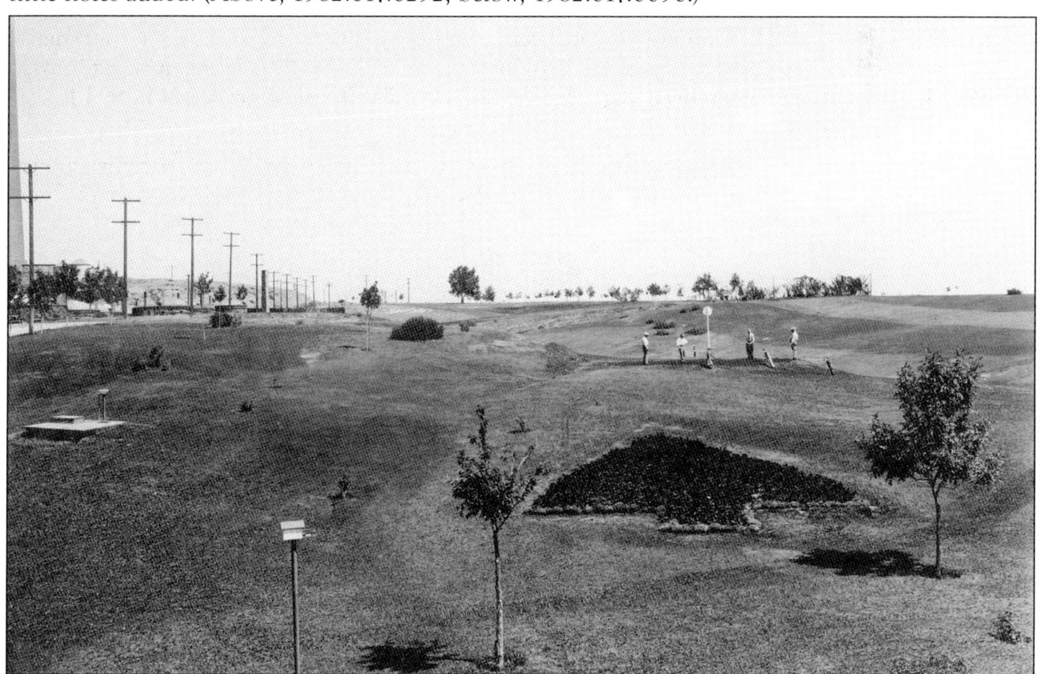

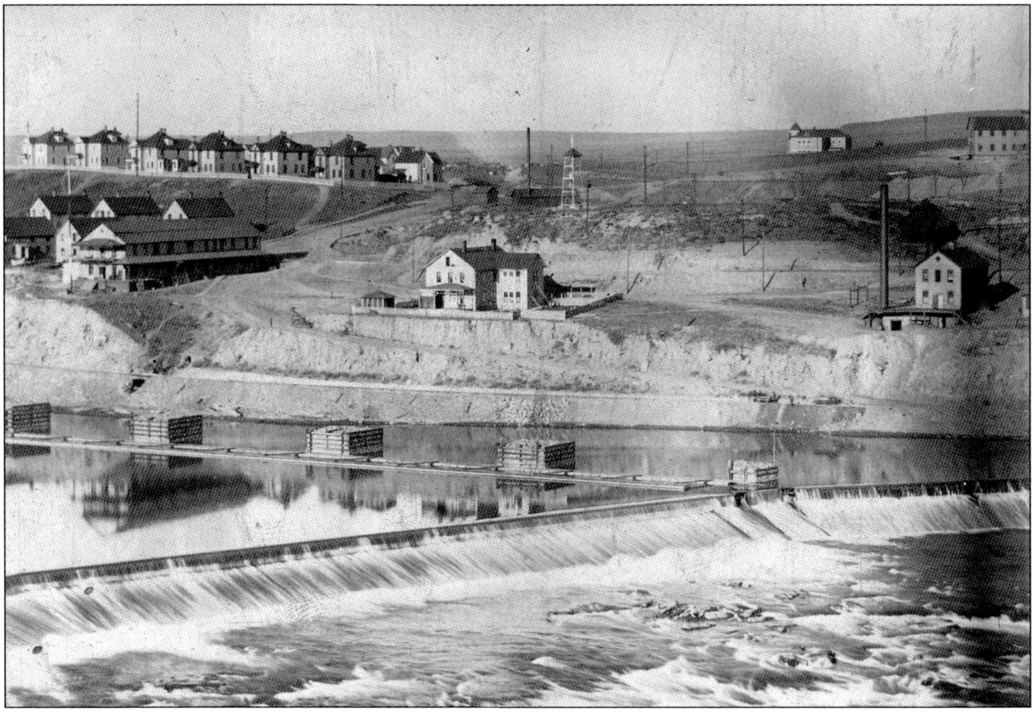

As mentioned in the first chapter, there were homes for the plant's management. Originally, there were 19 single homes and six duplexes. There was also a boardinghouse for single workers. In the photograph above is one of the earliest views of the homes, clubhouse, and boardinghouse. At the corner where the road turns from the old clubhouse toward the boardinghouse is a beautiful waterfall greeting visitors, seen below. Appropriately named Fire Bell Hill, the alarm bell for the fire brigade is perched above the waterfall in plain view, and later a small decorative chapel was added. (Above, 2009.020.0008; below, 1982.017.0312.)

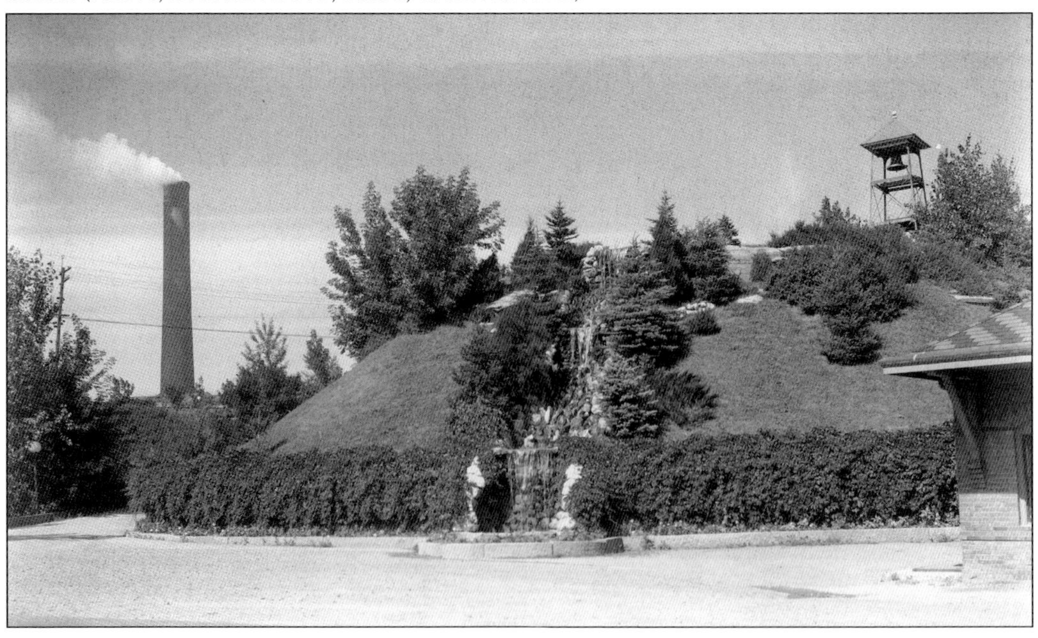

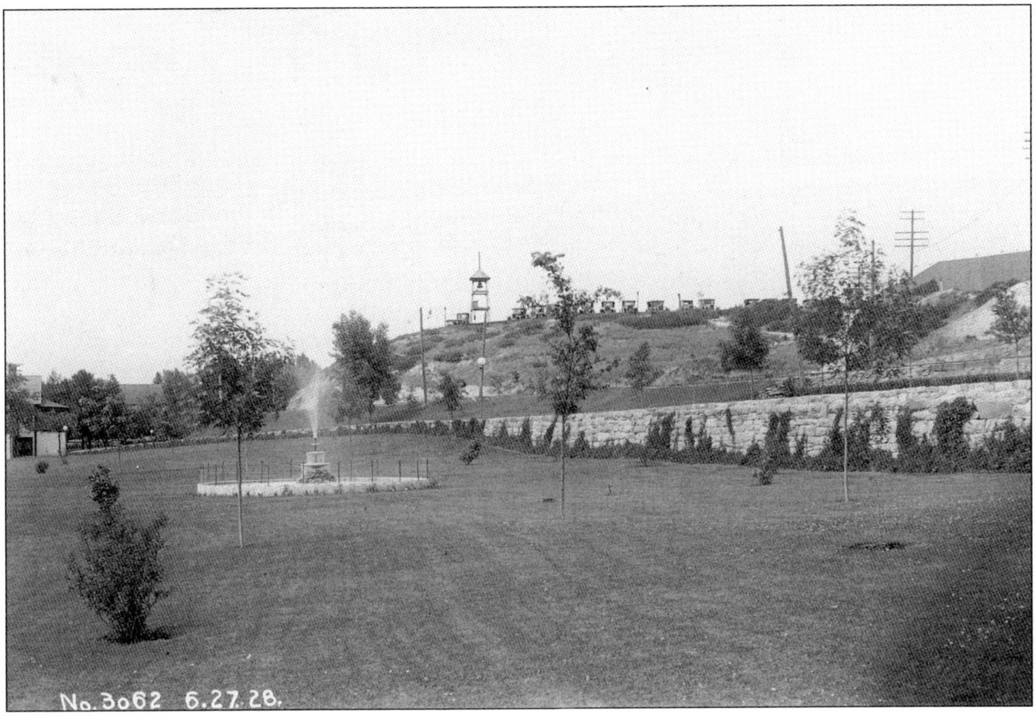

No. 3062 6.27.28.

The grounds closest to the housing were meticulously well kept, offering beautiful park spaces for families to enjoy. The photograph above gives a lovely view of Fire Bell Hill looking to the west. The push for beautifying the grounds and adding all the wonderful employee amenities came from the general superintendent of the plant in 1919, Albert E. Wiggin. He sought to promote athletic and recreational activities among his plant employees. It was under his tenure that the swimming pool, shown below, and the company club were built. (Above, 1982.017.0147; below, 1982.017.0129.)

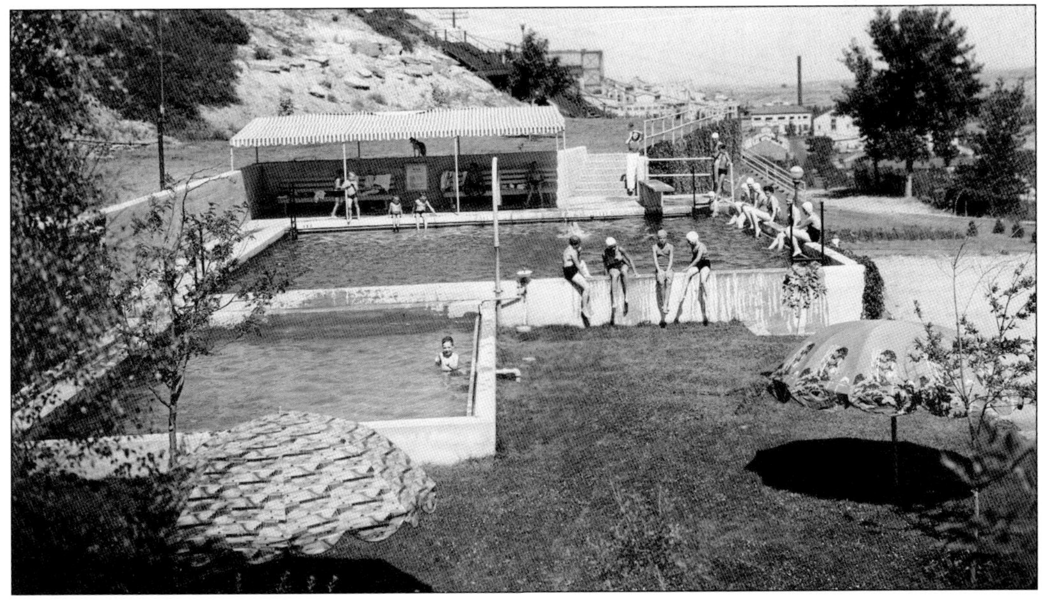

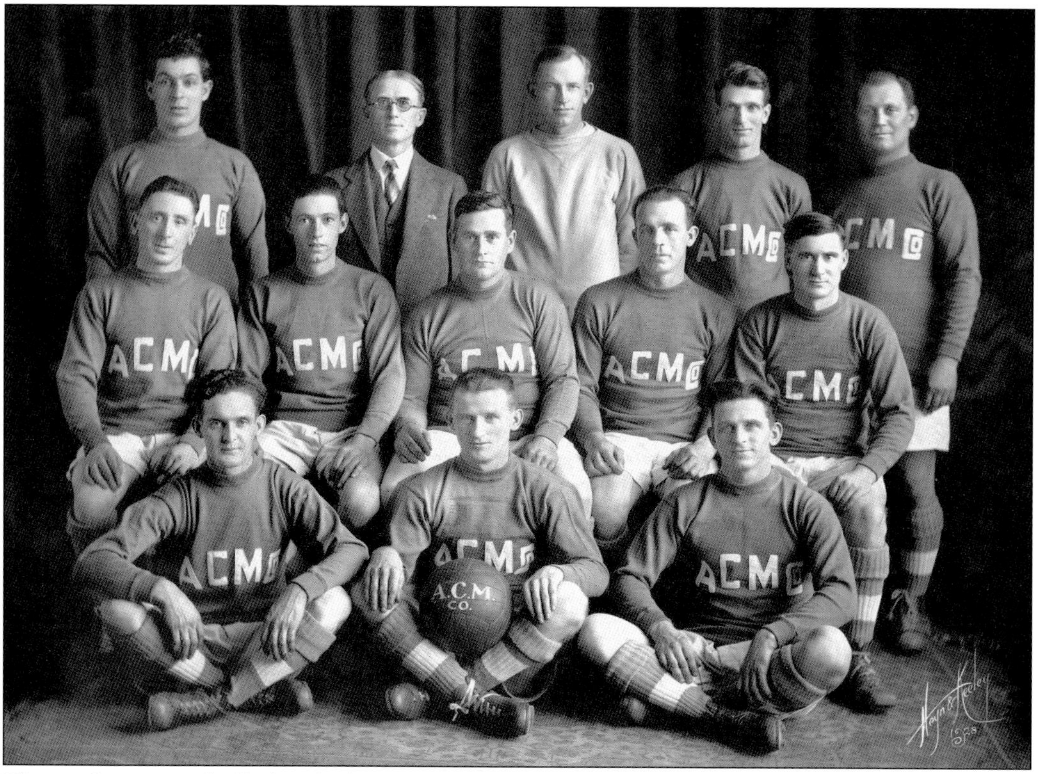

The workers were thrilled with the option to participate in sports and had teams for everything: bowling, soccer, softball, baseball, basketball, and hockey. Some sports, like soccer, only had teams for a couple of years. The team shown above represented the plant in a four-team city soccer league from 1928 to 1931. Interest dwindled, and the Depression caused the league to be disbanded. Other sports, like bowling, as seen below, became a mainstay of continuous entertainment. (Above, CCHS 197.0001; below, 1982.017.0204.)

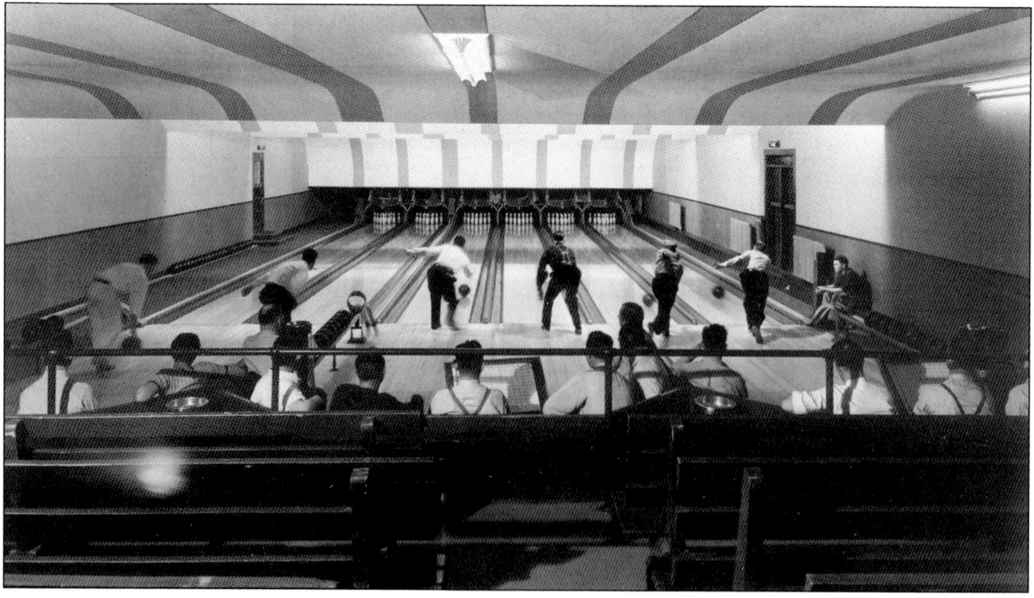

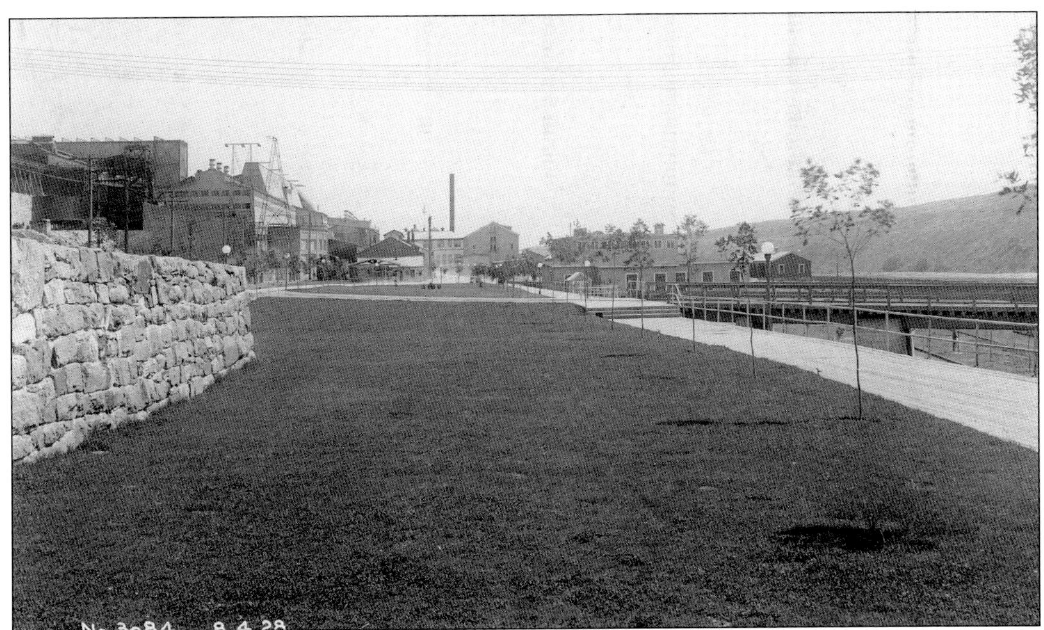

Continuing down the Low Line Road, as seen above, on the back left is the Zinc Roasting Building. To the far back right is a changing house that also housed the Sampling Department, the Low Line First Aid Station, and the Main Time Office. The group of buildings in the center includes the Blacksmith Shop, Machine Shop, and Main Warehouse. Past all of that, in the middle of the shop structures is the General Office Building, shown below, with a lovely lily pond fountain in front of it. (Above, 1982.017.0148; below, 1982.017.0310.)

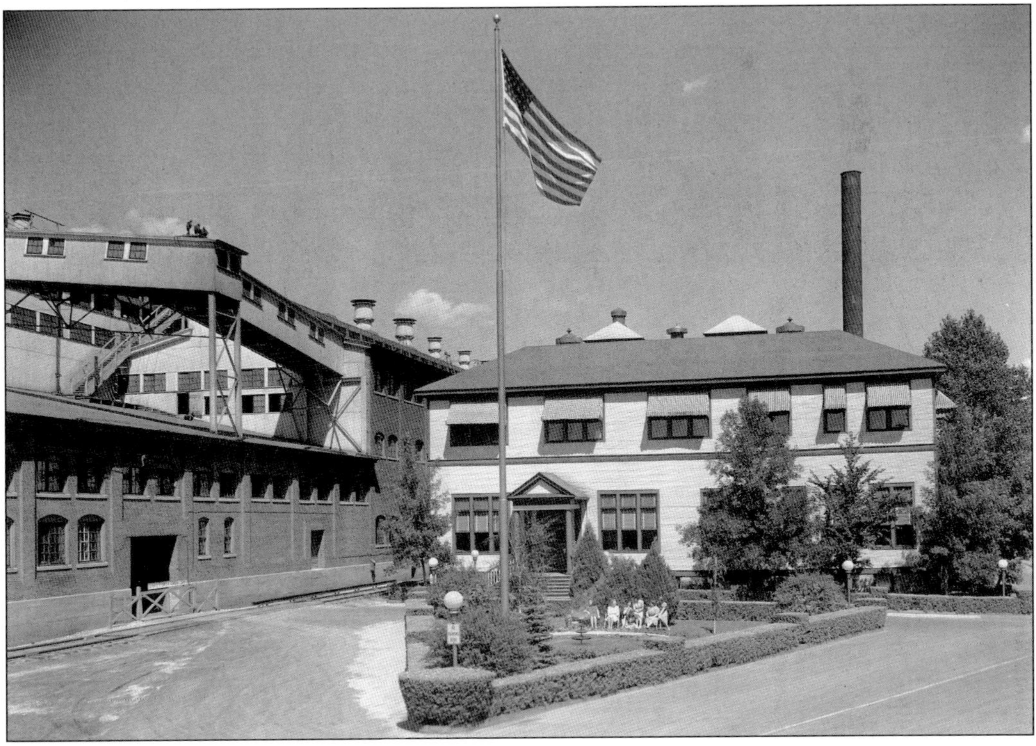

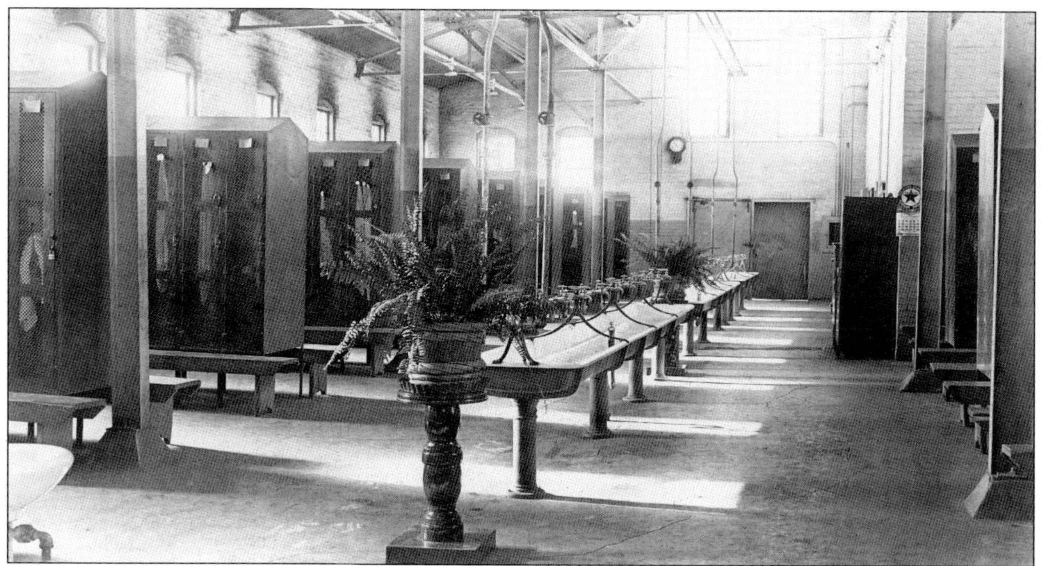

Dotted around the plant grounds were four change houses and five commissaries for the employees' convenience; an interior view of one of these is shown above. There were several first aid stations as well, and first aid became not only a large part of the training required but also a point of pride for the different departments. A first aid contest was initiated in 1919, and in 1923, the team from the plant won the world championship in Salt Lake City, as shown below. On the team were Gordon Gillis, Viggo Paulson, Joseph Macure, George Roberts, Edward Egan, and L.J. Deranleau. (Above, 1982.017.0067; below, 2022.027.0006.)

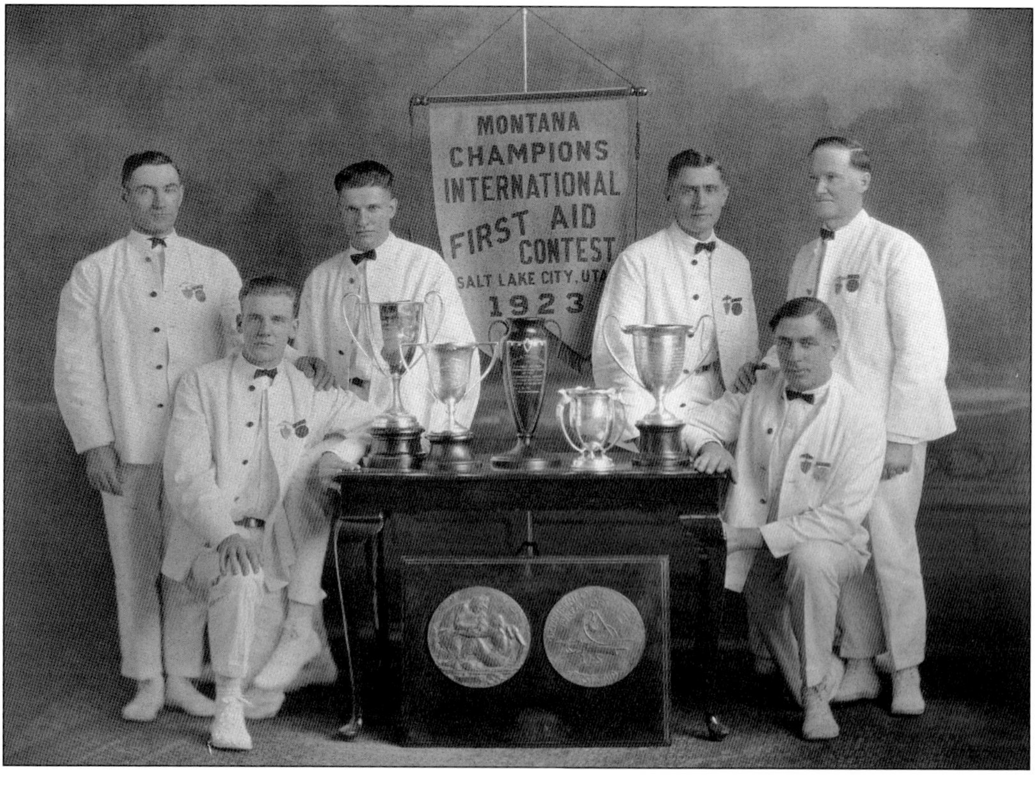

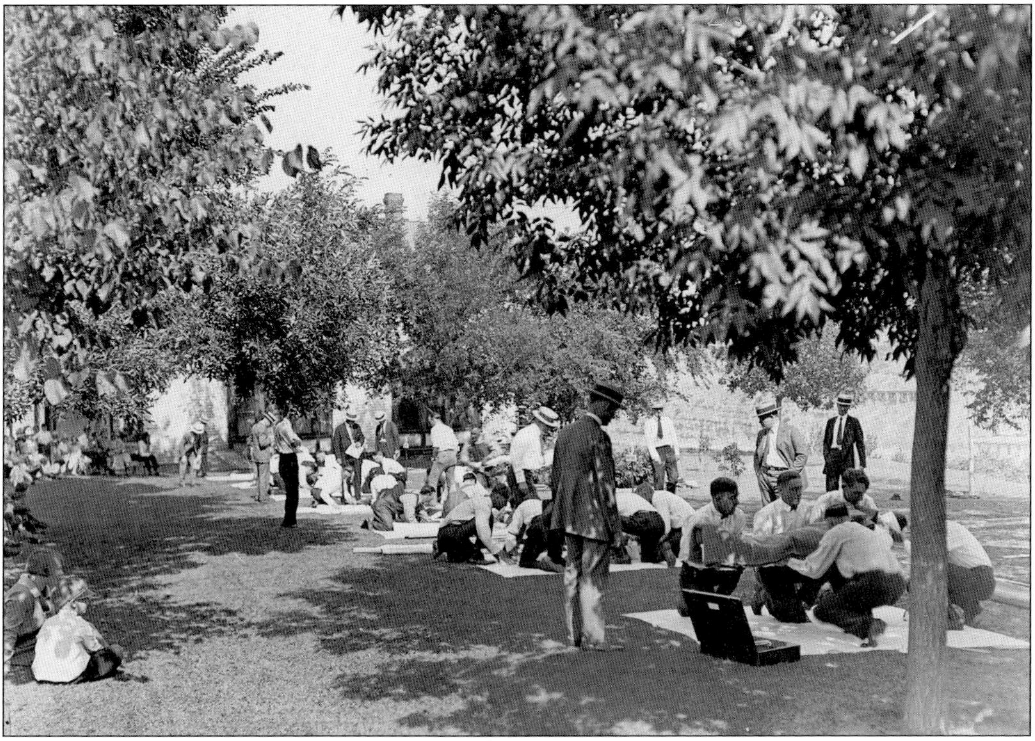

The first aid contests became a yearly competition that evolved into family days with picnics and other games, as seen above. Safety was always at the forefront of Anaconda's priorities, and several other sorts of incentives were put in place. For every 10,000 days of no accidents, departments would get free cigarettes, as shown on the sign in the image at right. The clubhouse would keep track of the brand that each man liked to smoke. (Above, 1982.017.0136; right, from 2015.089.1.)

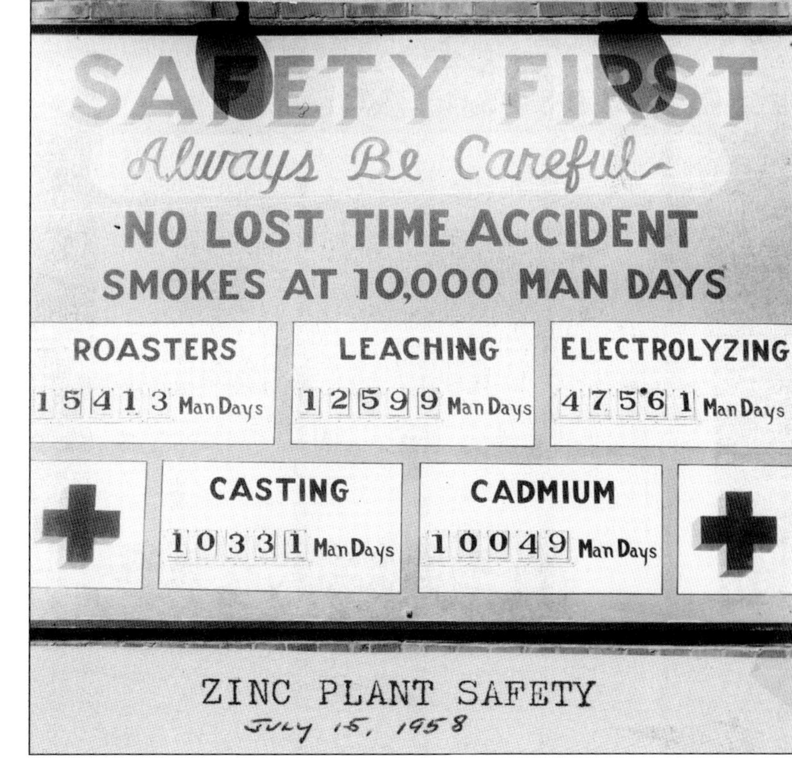

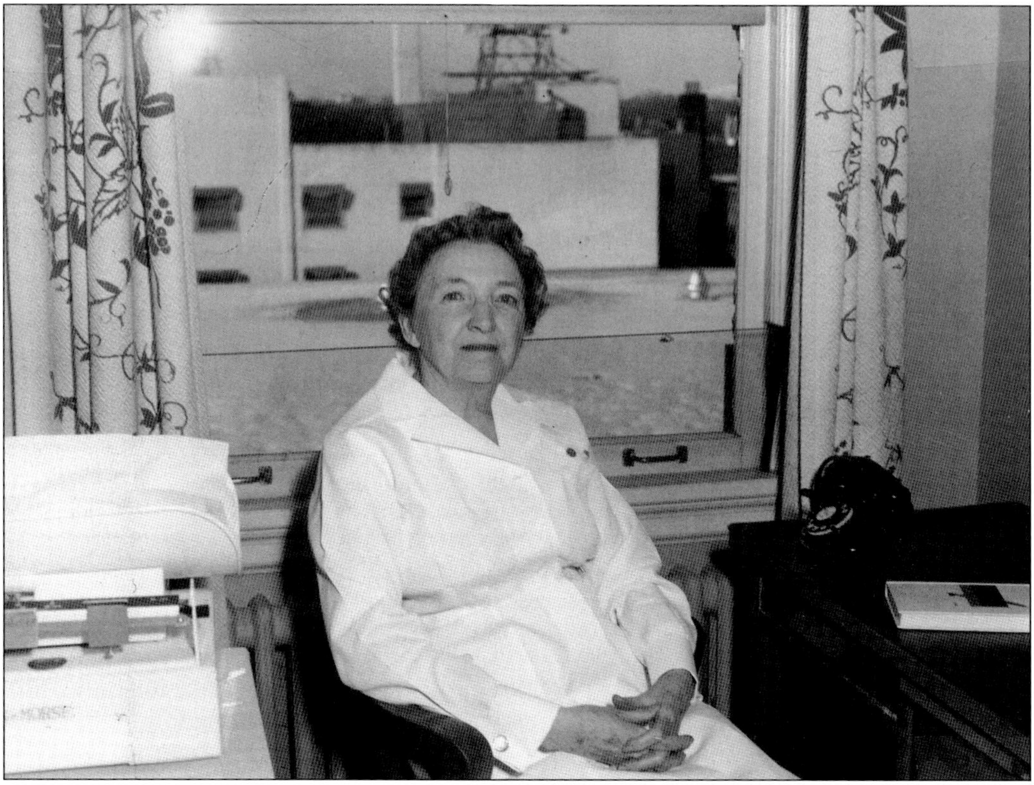

The children were given special care and attention. In 1926, the medical staff of the plant acquired a suite of offices in the First National Bank Building in downtown Great Falls. At that time, Clara Dahl was added to the staff. She graduated from the Montana Deaconess Nursing School, and her specialty was the care of babies. She made house calls, checked infants, and provided instruction for proper care of children as well as individual recovery. Above is a photograph from her later years with the company. The children would then go on to attend Hawthorne School, seen below. The land for the school was leased from the Boston and Montana Company and was first named the Copper Smelter School in 1897. (Above, from 2015.089.2; below, 2022.005A.0485.)

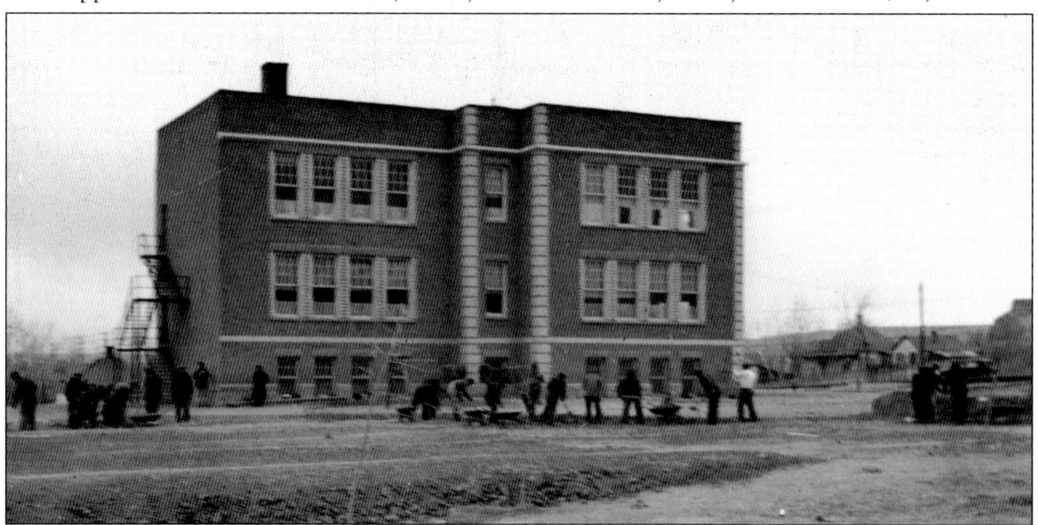

A big part of working at the company was the unions that represented each of the professions. The largest one was the Mill and Smeltermen's Union Local No. 16. The unions kept the company in check and worked to get the workers better conditions and pay. They were also a source of support for their fellow workers in other cities across the state. There was always a representative at the plant that workers could go to, like Ray Graham, featured at right. Many stayed with the company for decades. Alex McLeod, who worked for the Boston and Montana Smelter from 1892 to 1897, came back for a visit on September 19, 1958, featured below, and shared stories with the boys. (Both, from 2015.089.2.)

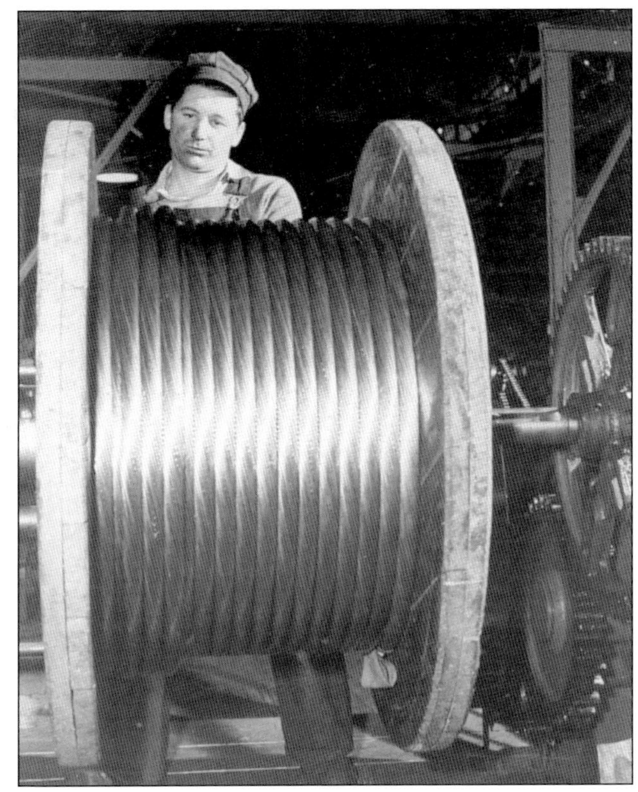

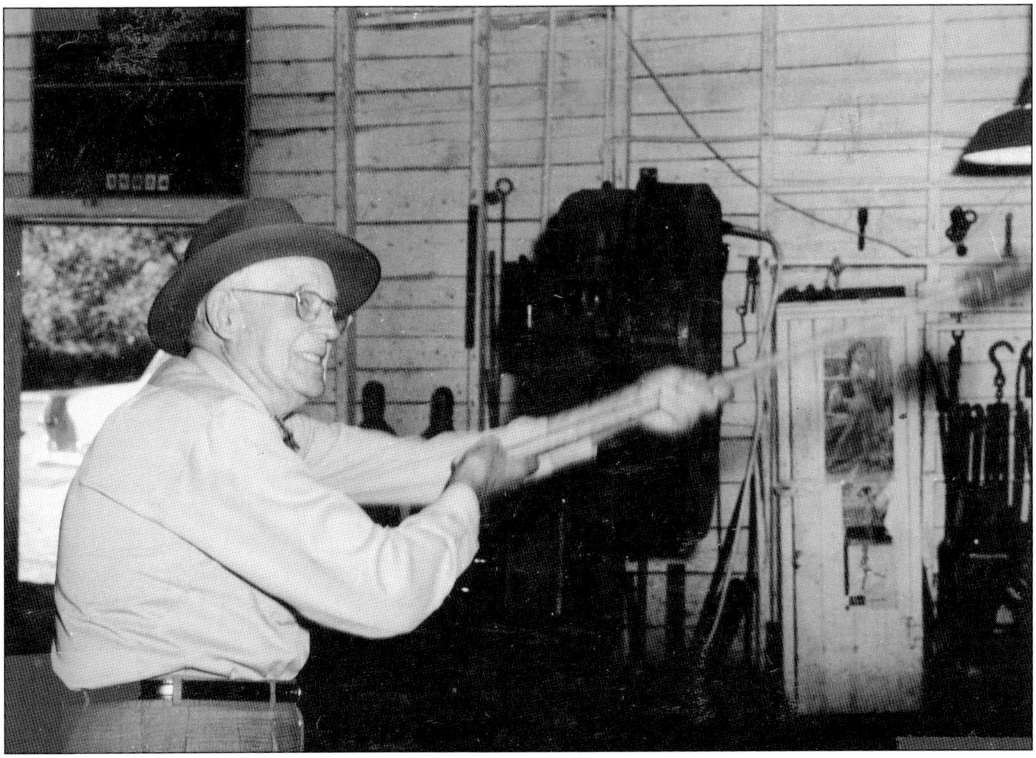

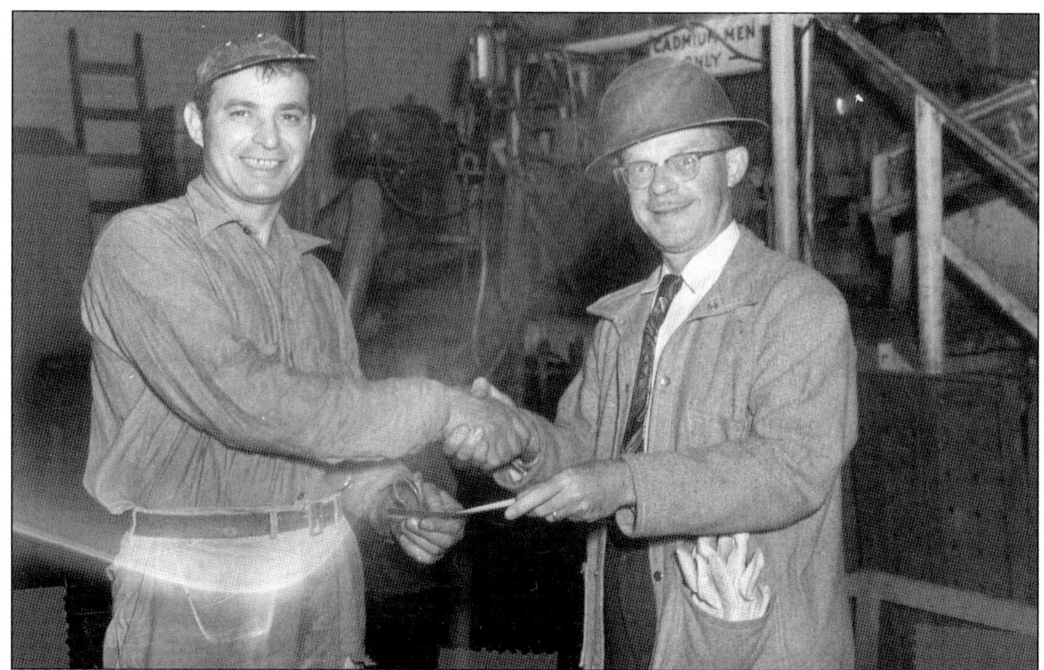

The Anaconda Company was proud of its employees and encouraged their growth and ideas. It established a suggestion award, giving a cash award for ideas that improved safety and the processes at the plant. Above, Wes Schnieder is receiving his award from Alfred I. Alf. Anaconda promoted from within and would put employees' achievements on display in the newspaper frequently. When it was time for retirement, the company celebrated their bodies of work, like with Charles Dohnmeyer, featured with his retirement gifts below. A good example was Alfred I. Alf, a metallurgist in the Zinc Department, who during his career received three patents on processing and recovery of indium and germanium. (Above, from 2015.089.2; below, 1982.017.0237A.)

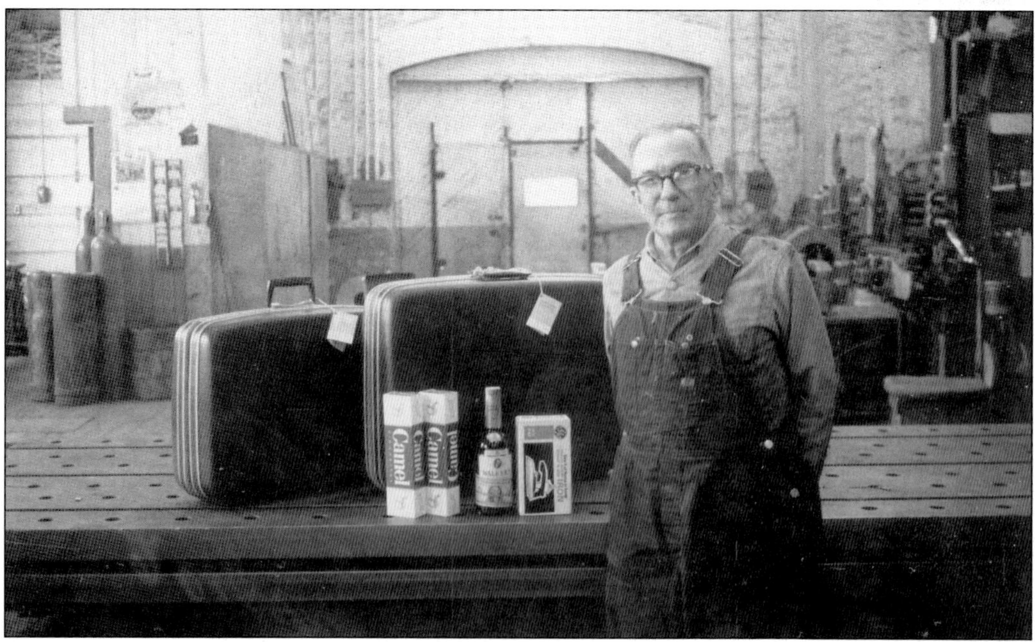

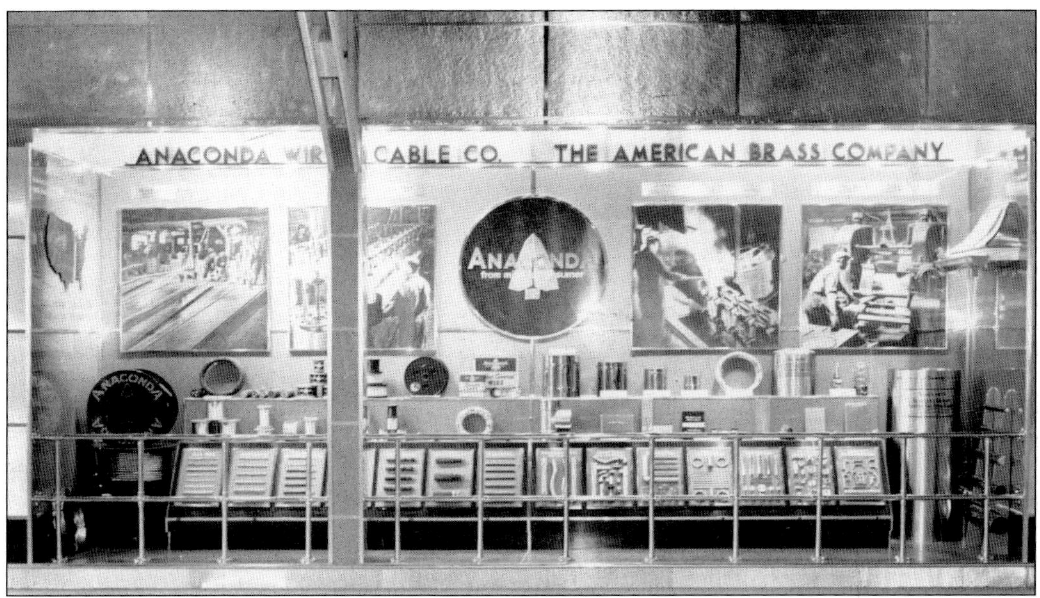

Every chance it could, the Anaconda Company would showcase the work being done, giving the public a better idea not only of how ingrained it was in everyday lives of those working for the company but also how the copper, zinc, and other elements were important to daily tasks in the home, at school, and for other professions. An exhibit featured in these photographs was put on display at the North Montana Fair in Great Falls in 1939 and gave an in-depth look at the processes, products, and uses of the products. (Above, from 2015.089.1; below, from 2015.089.2.)

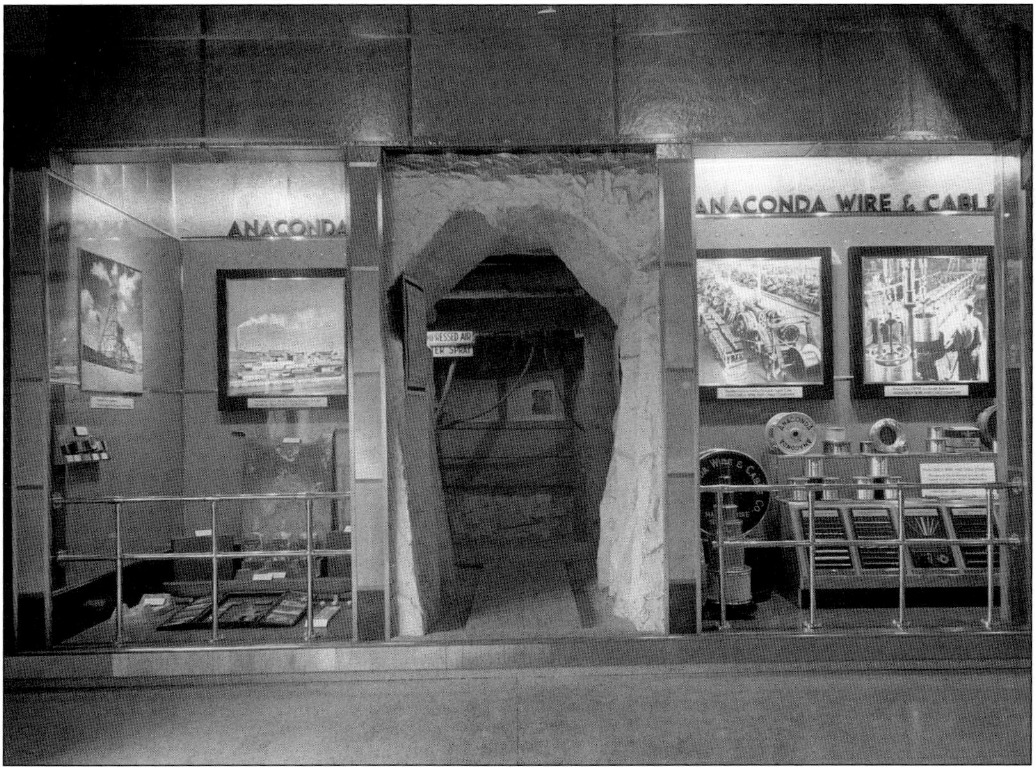

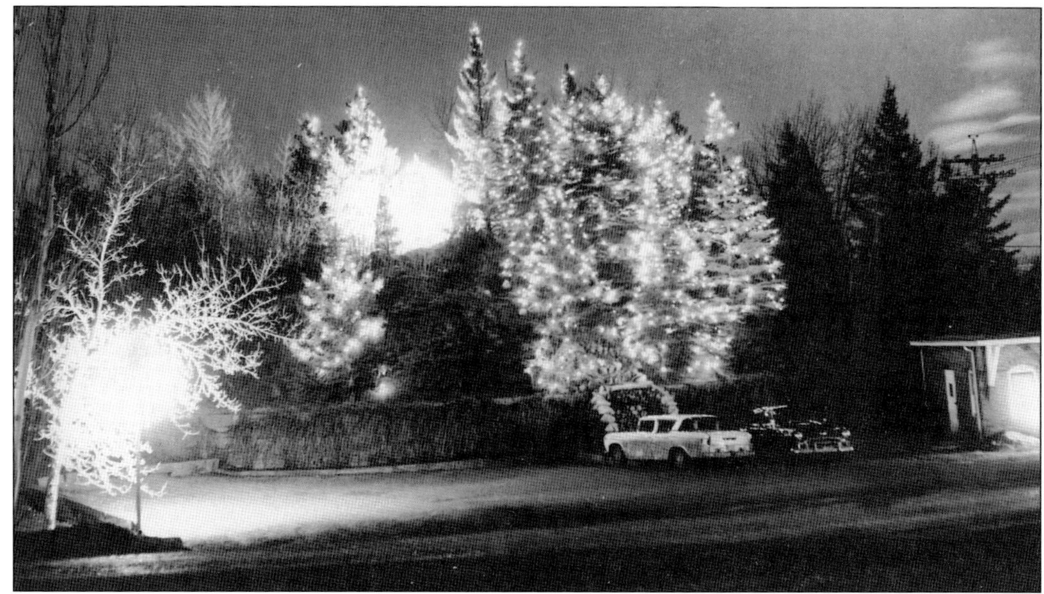

One of the most beloved memories of the surrounding communities was the Great Falls Plant at Christmastime. All the trees were decorated with lights, as seen in these photographs, and people were invited to drive through the grounds. This tradition began in 1920 with a single 40-foot tree from Monarch, Montana. It was decorated with 365 lamps, one for each day of the year. In 1932, additional trees began to be decorated as well. It gave the place an ethereal feeling. (Above, 1982.017.0108; below, 1982.017.0094.)

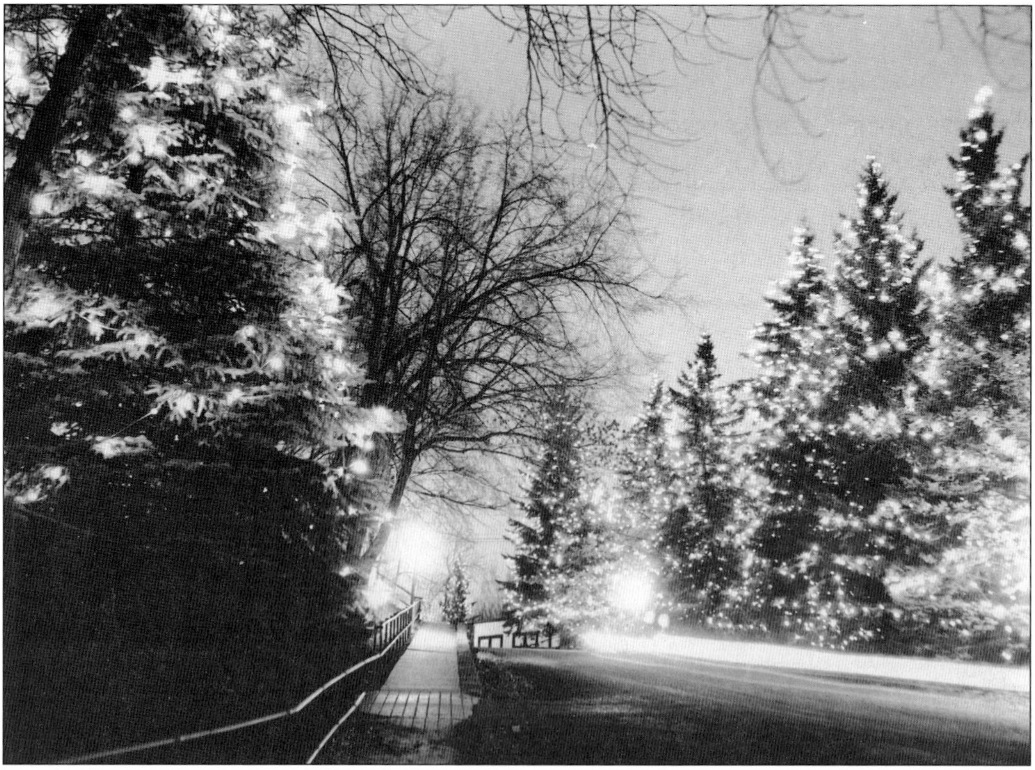

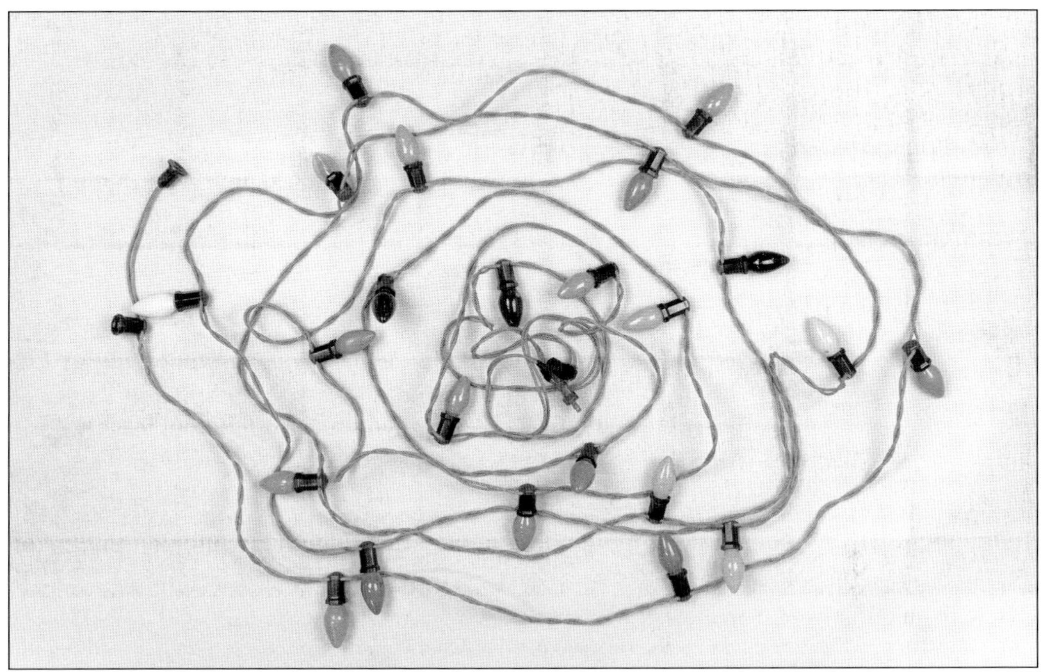

By 1940, there were 11,000 lights decorating Smelter Hill. During World War II, the lights were loaned to the Great Falls Army Air Base for its Christmas party, since they could not be put out on display due to blackout measures. After the war, more lights were added each year. In 1965, over 16,000 vehicles drove through to see the 115 trees decorated with over 28,000 lights. It was referred to as Fairyland, as is reflected in the image below. Above is a string of the lights used on Smelter Hill that are part of the collections at The History Museum & Research Center. (Above, 2015.088.1; below, 1984.164.0003.)

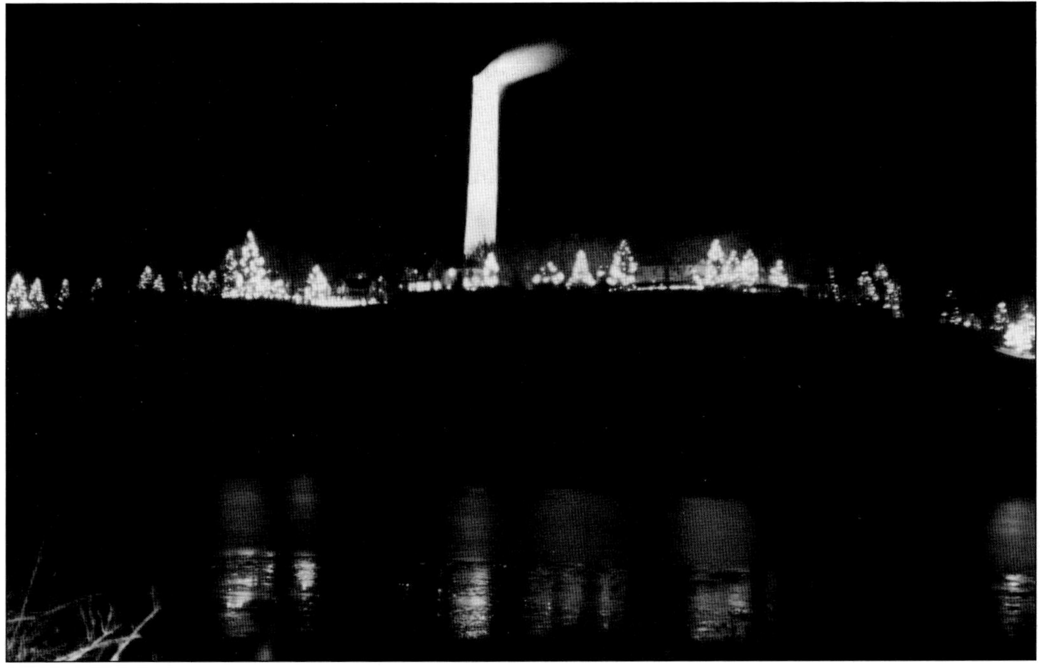

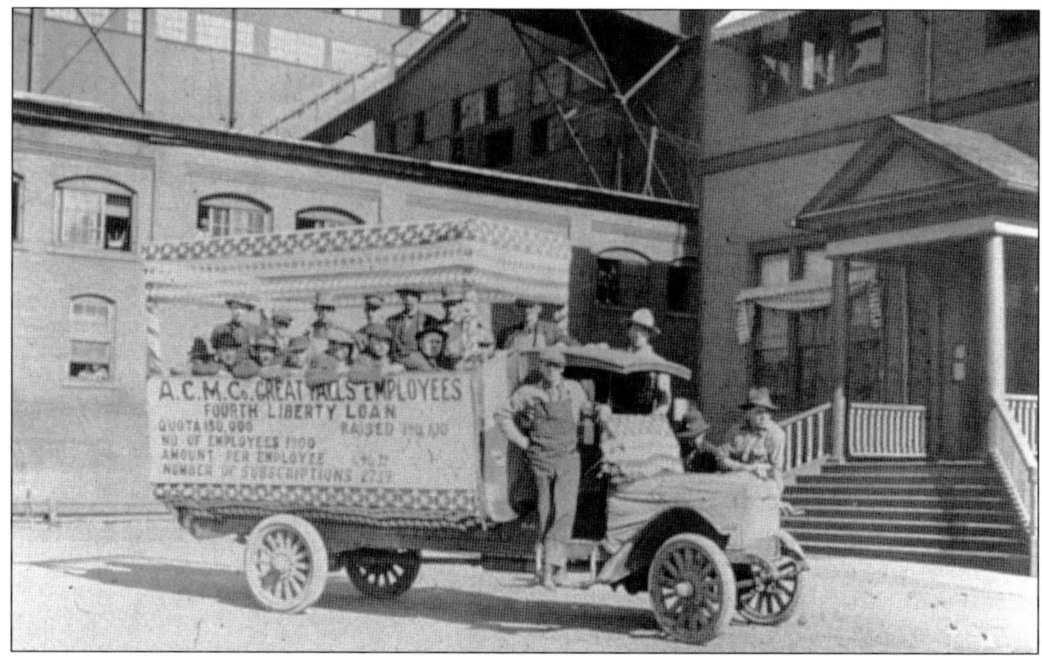

During wartime, the plant became indispensable. Copper and zinc were needed to make bullets, helmets, and weapons. Liberty bonds were heavily encouraged, as seen above, and those who did not go to war worked hard for their friends and family who were fighting for them. During World War I, 351 employees served, and eight lost their lives: William T. Beacom, Jack F. Flannery Jr., Carl L. Furstenau, Joseph P. Lyons, Erwin E. Marvin, Patrick E. O'Neill, Harry A. Smith, and Gustaf Winberg. Their names were inscribed on a bronze tablet mounted on a boulder in front of the General Office Building, as seen in the image below. (Above, 1996.098.0023; below, 1982.017.0311.)

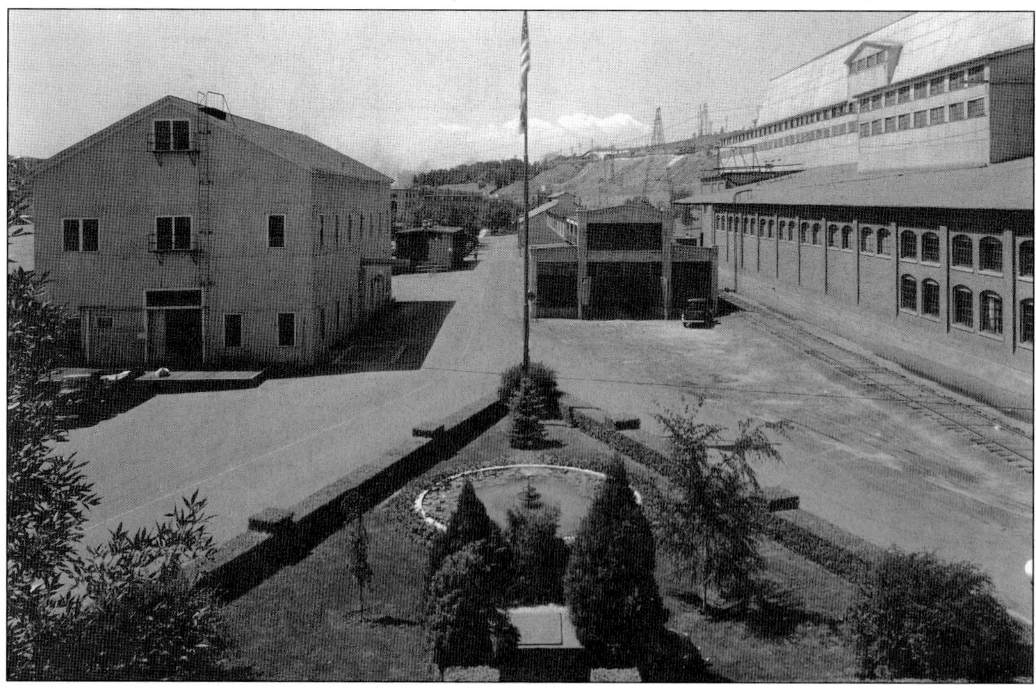

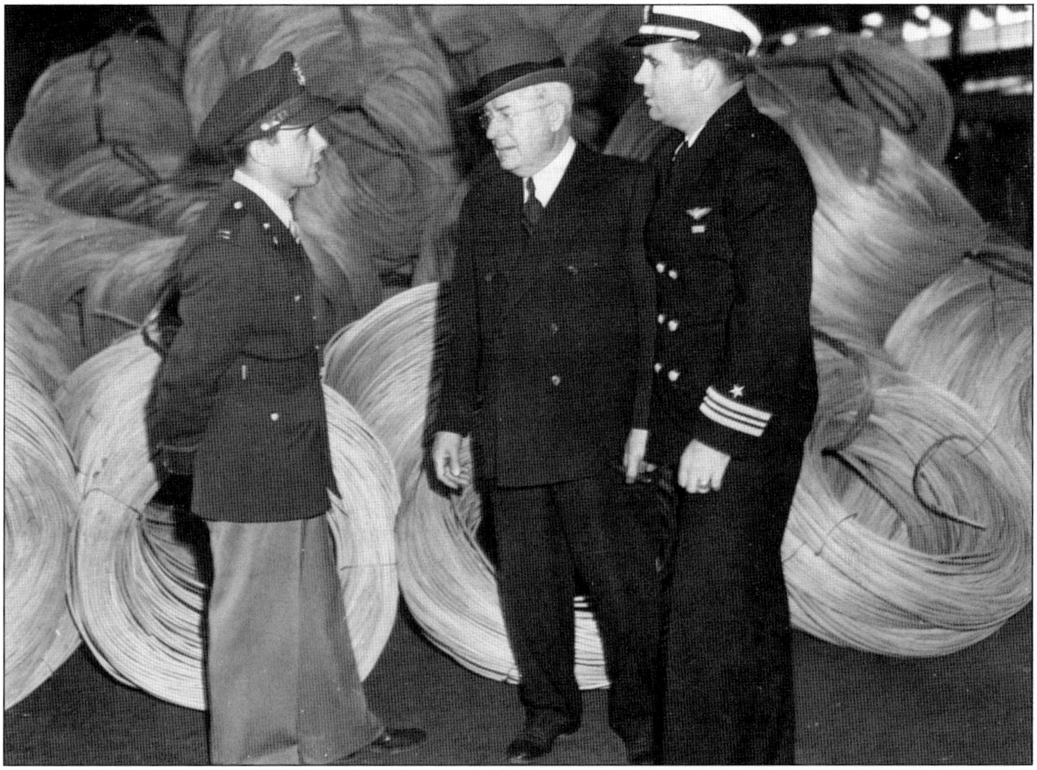

When war was declared again in December 1941, the employees of the Anaconda Company across the state of Montana were ready to do their part. On June 3, 1942, Lt. Comdr. Edward H. O'Hare, US Navy (above, right), and Capt. Hewett T. Wheless, US Army Air Corps (above, left), paid the Great Falls Plant a visit. They are speaking with D.M. Kelly (above, center), vice president of the Anaconda Company. Captain Wheless was the pilot and commander of a Flying Fortress that was attacked by 18 Japanese planes and successfully returned after bombing his objective. Lieutenant Commander O'Hare was a fighter pilot who shot down five Japanese bombers in succession. They both spoke to the need for copper and zinc for the war effort. At right, Maj. James R. Baker (left), who accompanied the two heroes, is seen speaking with F.S. Weimer (right), general superintendent at the Great Falls Plant. (Above, from 2013.080.2; right, 2014.018.0007.)

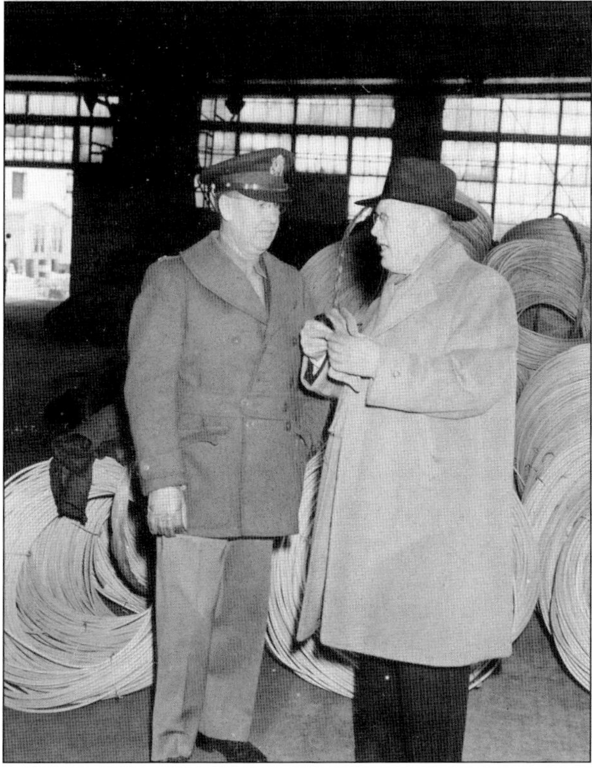

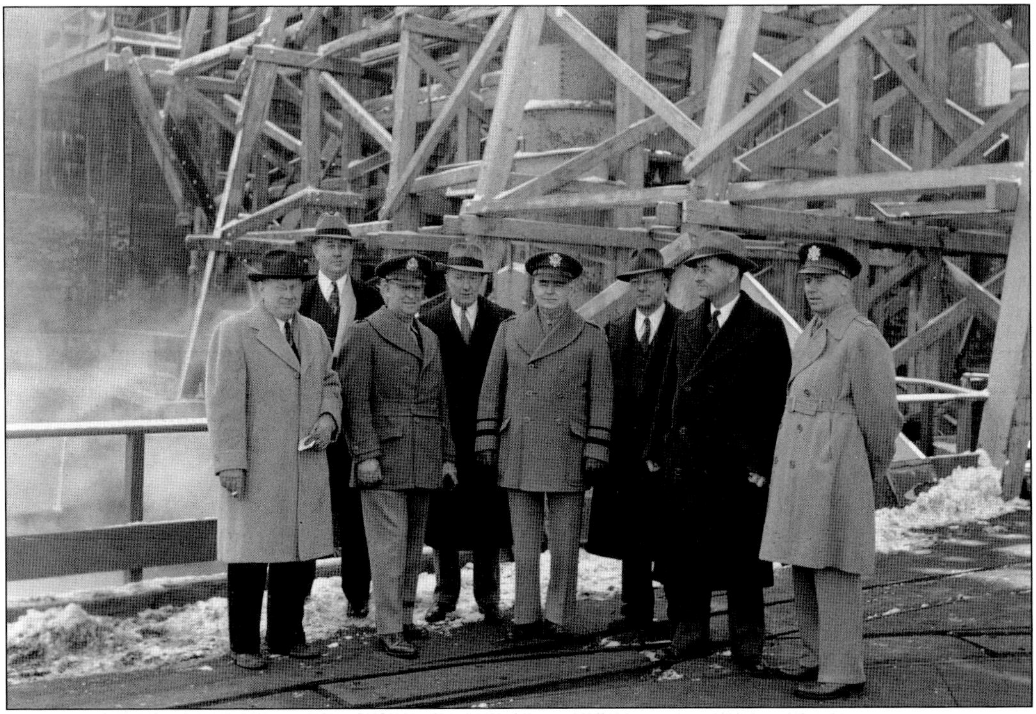

During World War II, the Great Falls Reduction Department and the Wire and Cable Company were awarded five consecutive Army-Navy "E" Awards for excellence and efficiency. Officials and military personnel admired the work at the Great Falls Reduction Department, as shown above. The Great Falls facilities were awarded the "E" flag, flying in the image below, and each consecutive year, a star was added to the flag. From 1941 to 1944, the plant produced 1.2 billion pounds of zinc, 1 billion pounds of copper, and 7.5 million pounds of cadmium. Rarer metals like indium and gallium were also produced in large quantities. Indium was used to plate airplane engine bearings to prevent acid wear. (Above, 2014.018.0004; below, from 2015.089.1.)

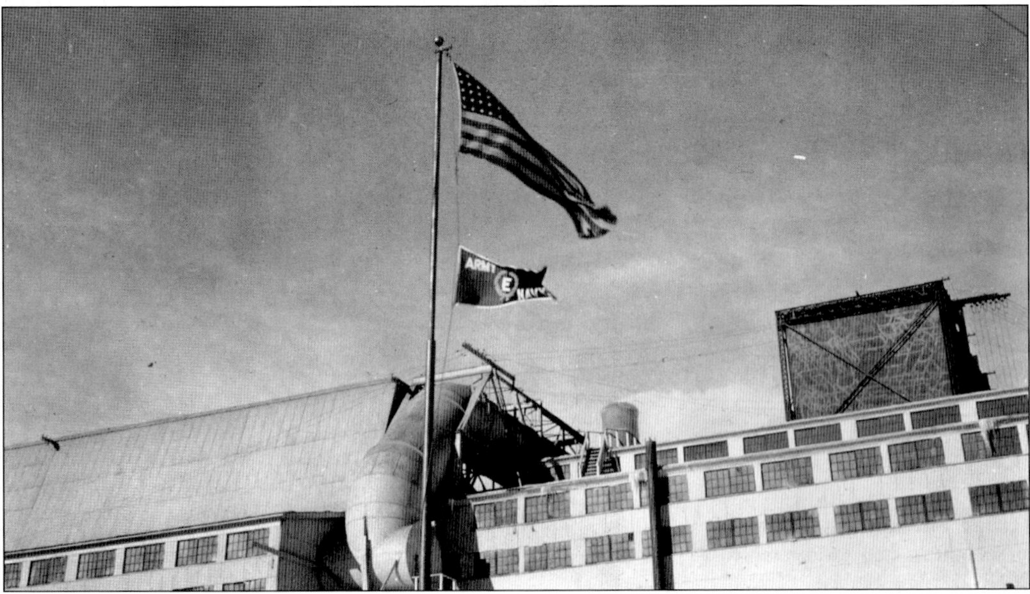

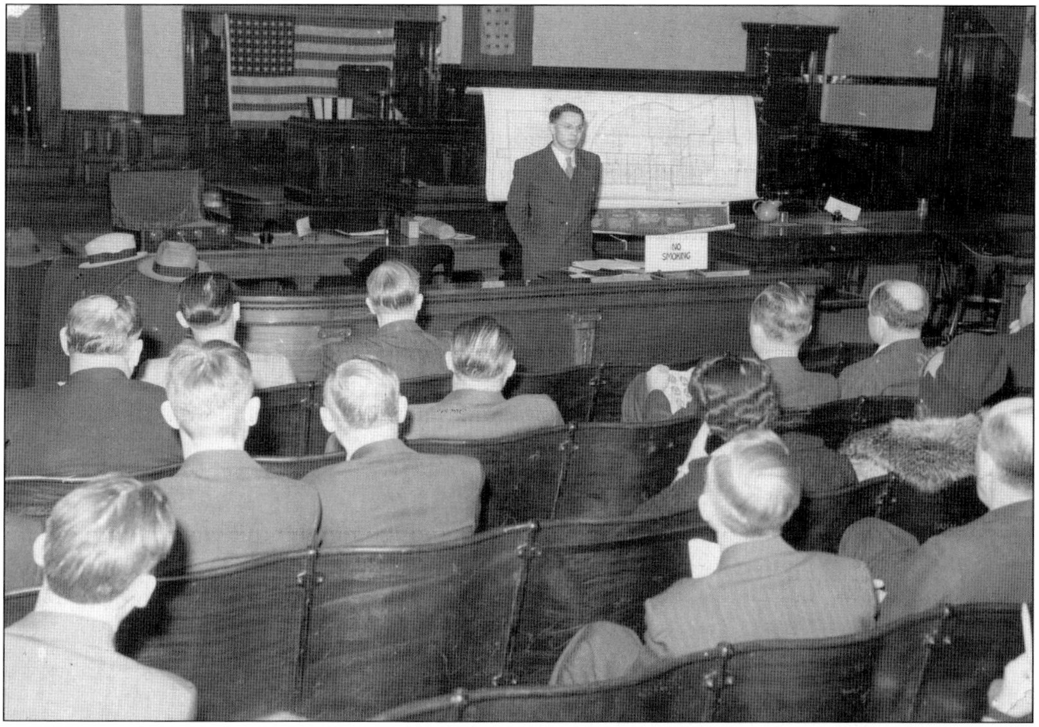

Besides Liberty bonds, employees found other ways to contribute in their communities, like participating in civil defense and Red Cross training. Albert French, above, served as commander of the Citizens' Defense Corps for Cascade County. At the end of the war, 544 employees from the Great Falls Plant had served in an armed service, 17 employees were killed in action, and 3 were reported missing in action: Edward J. Angermeier, Louis Blumfield, William J. Burns, Edward D. Clark, Harold M. Helgeson, Warren J. Hines, Frank J. Krsul, James J. Matteucci, Francis K. Ringler, John A. Saxbury Jr., Kenneth I. Scott, Donald H. Siegling, Jacob T. Tabarracci, Garnet F. Van Zant, Louis M. Weber, John Wick, Roy A. Wright, George S. Horan, Earl Ramsted, and Edward J. Seffani. When World War II ended, the plant put out a celebratory newsletter, as seen at right. (Above, from 2015.089.2; right, 1988.129.4D.)

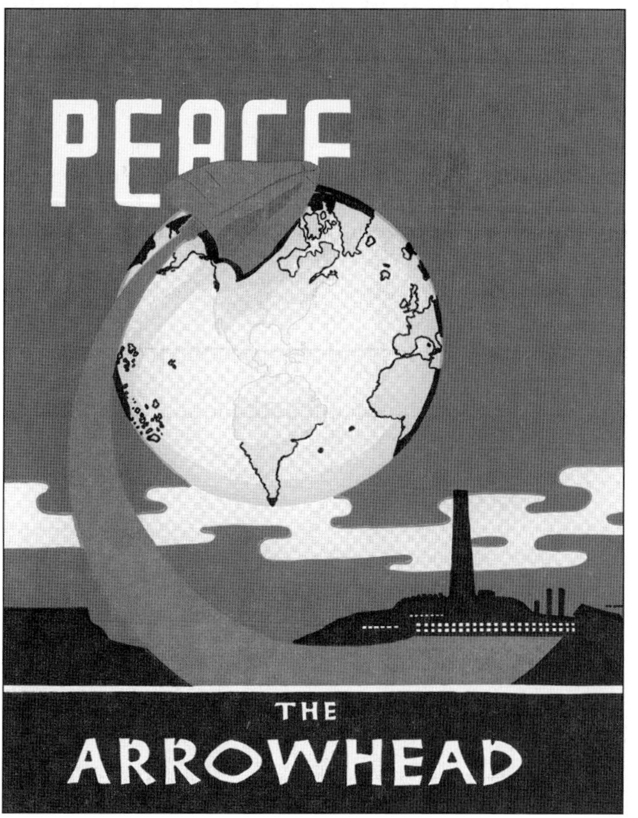

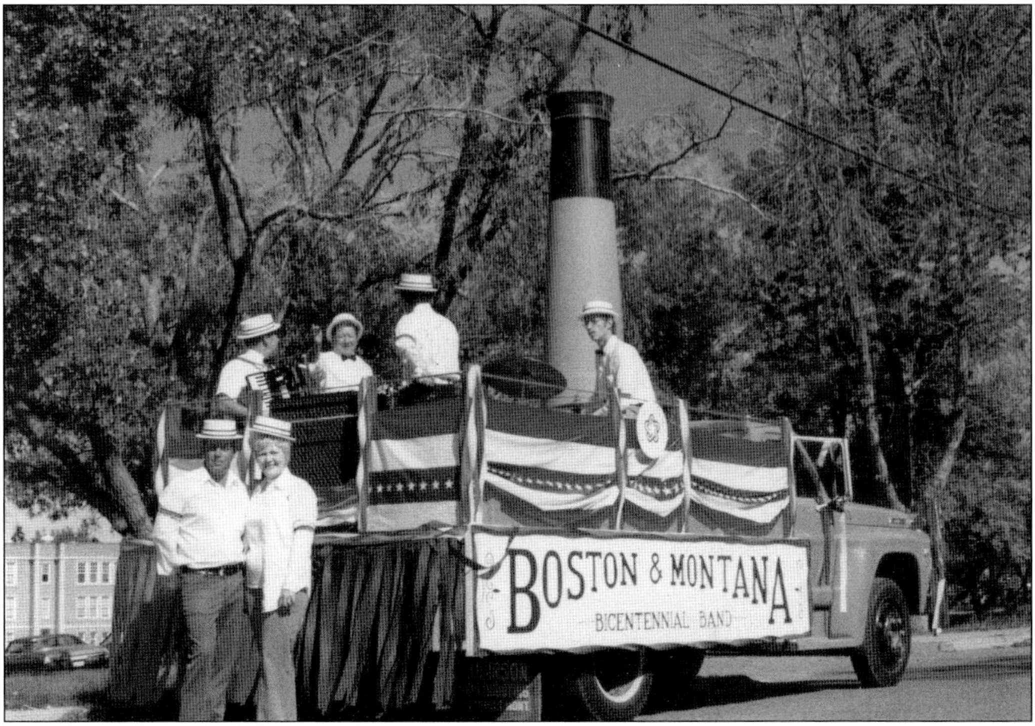

The employees of the Great Falls Reduction Department Plant were truly proud of their work and the company. Multiple generations would work for the company at one time; they celebrated good times and sad times with their fellow employees and supported each other. The Anaconda Company was part of every aspect of their lives. Featured here are views of the parade float the employees made for the 1976 bicentennial parade in Black Eagle. (Above, 1982.017.0493A; below, 1982.017.0495B.)

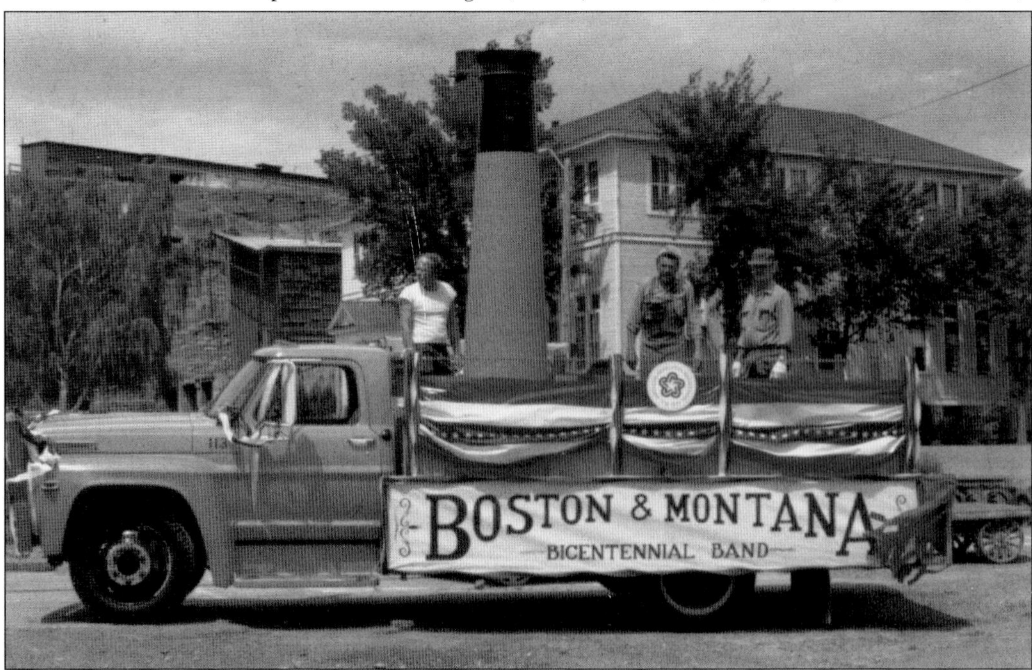

Six

The Long Goodbye

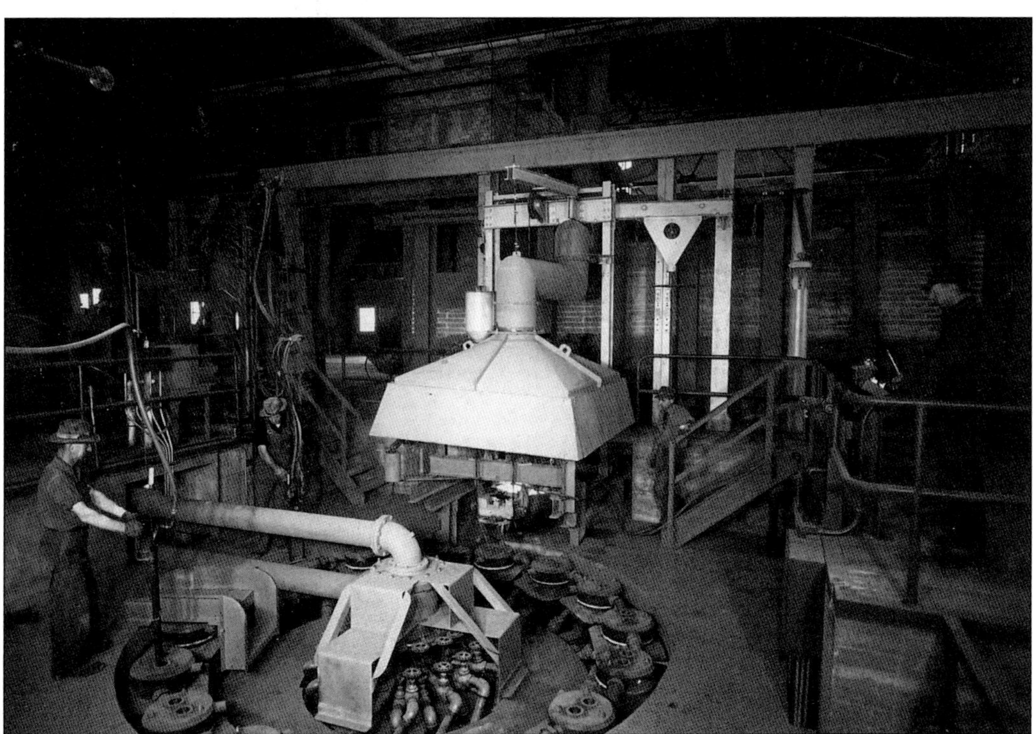

After World War II ended, the demand for copper and zinc declined. The Anaconda Company found new ways to expand and increase production with the Aluminum Plant in Columbia Falls, Montana, which in turn added aluminum processing to the Great Falls Wire and Cable Mill. The Great Falls Plant also added billet shapes for copper to produce copper pipes. Billets are cast vertically, making five-foot-long cylindrical bars that are three inches in diameter. The casting machine can be seen above. (2023.043.64.)

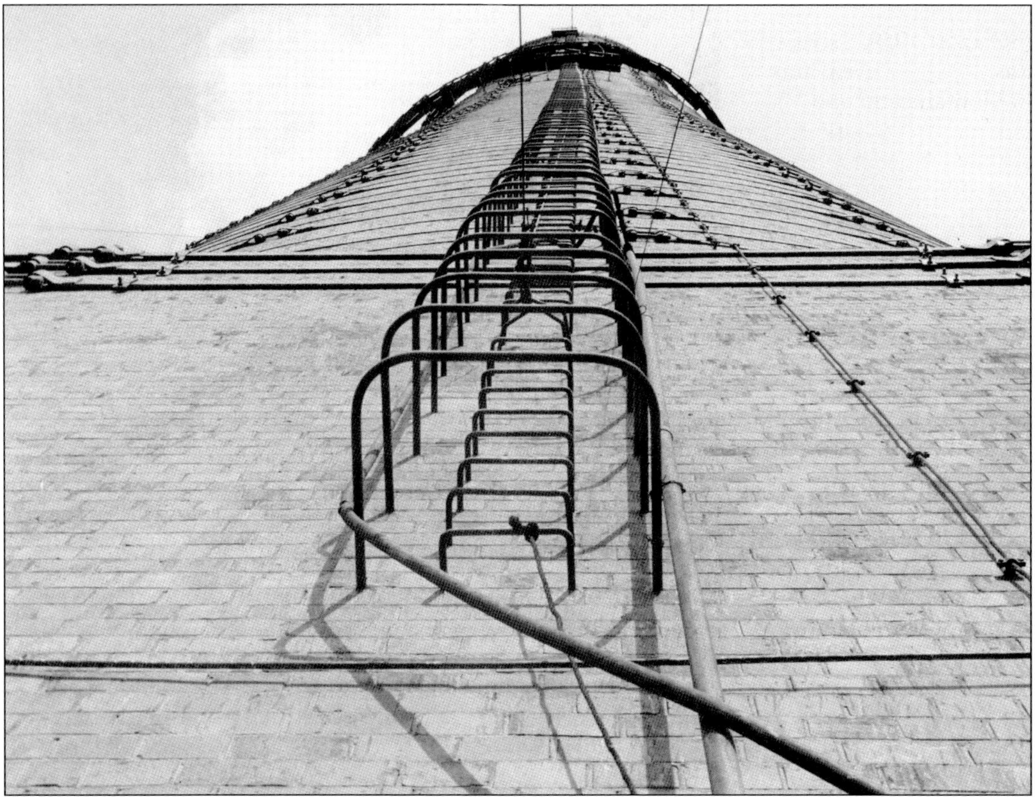

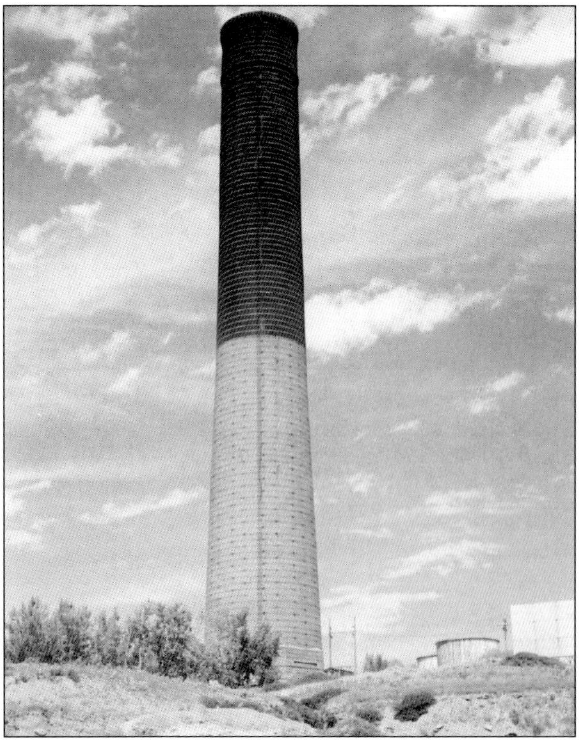

Over the summers of 1955 and 1956, workers from Custodis Construction Company (previously Alphonse Custodis Co.) were called upon to do exterior preventative maintenance to the Big Stack. They placed 118 steel bands on the stack for structural support, and cracks were repaired, as seen above. In 1957, the top 150 feet of the stack was painted black to help seal against moisture, as seen at left. (Above, 1982.017.0283; left, 1995.071.0052G.)

In August 1959, a strike was called for five plants in Montana and one in Utah over insurance disputes. However, there were still some repairs that needed to be done on the Big Stack. Damage to the interior can be seen at right and below. The communal importance of the Big Stack made its upkeep separate from union issues. The interior repairs were the first time the inside of the stack got a cleaning since it began its work. The cleaning was done by personnel of the Custodis Construction Company with help from the local unions. The strike lasted 177 days. (Both, from 2015.089.1.)

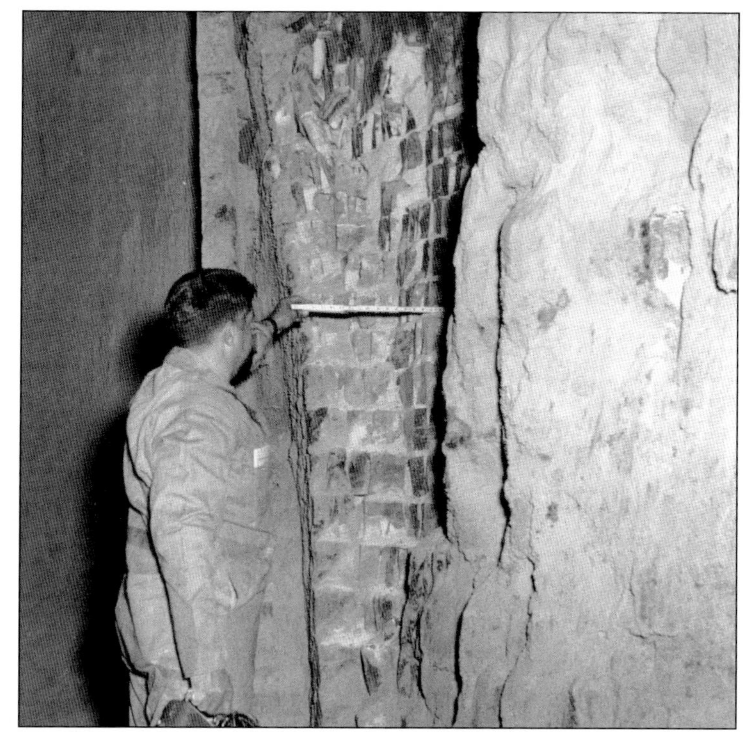

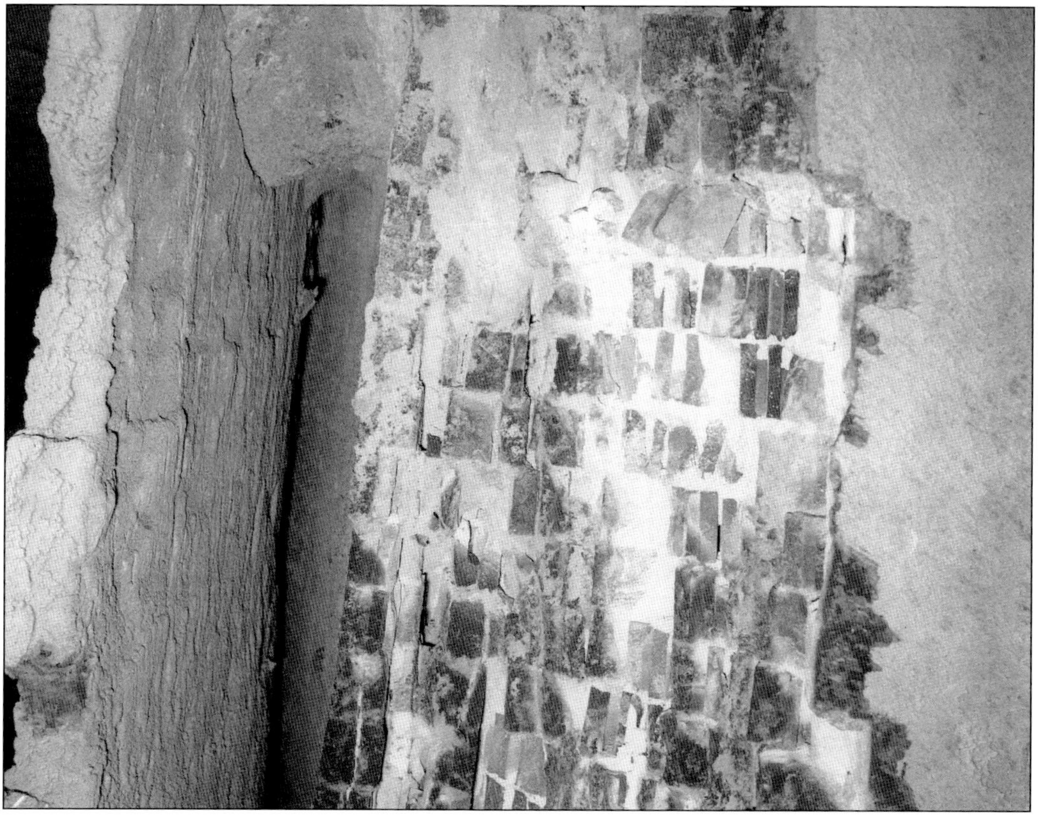

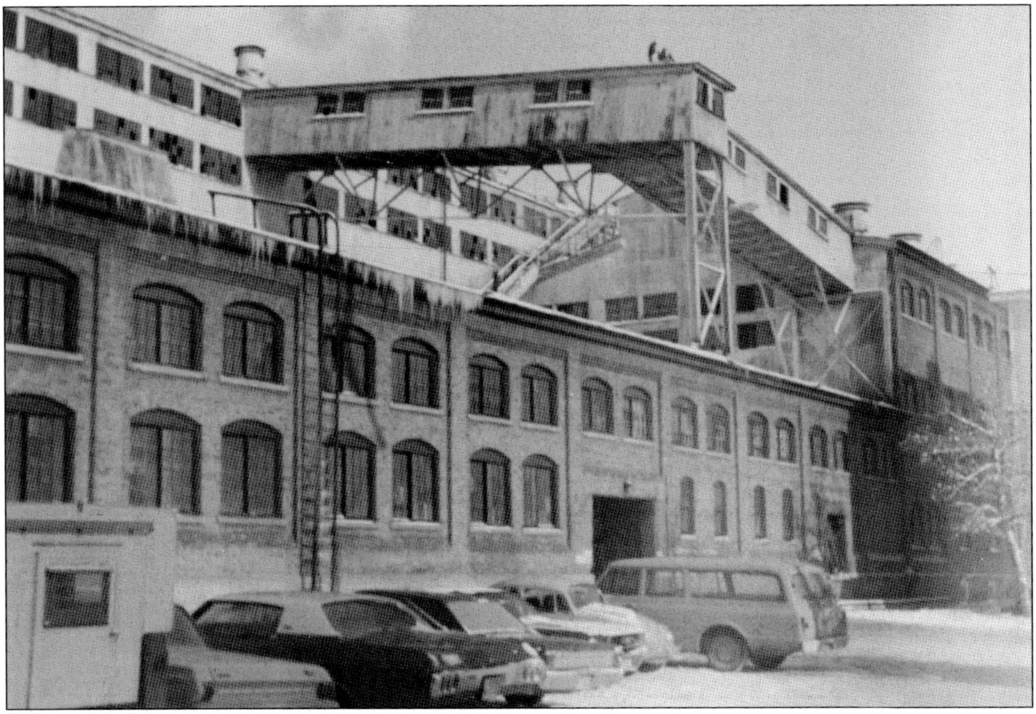

Strikes continued to plague the Great Falls Plant over wage disagreements, automation reducing jobs, or insurance coverage. Shortages caused temporary closures, but demand never returned to wartime levels. In 1969, the Zinc Plant in Anaconda, Montana, was closed, and the work was transferred to the Great Falls Plant. This was the start of Anaconda's decline. In 1972, a survey was done at the Great Falls Plant assessing the value of the buildings. Above is part of the flue furnace substation building, and below is the Furnace Refinery Building. (Above, 1992.086.0002-47; below, 1992.086.0001-17.)

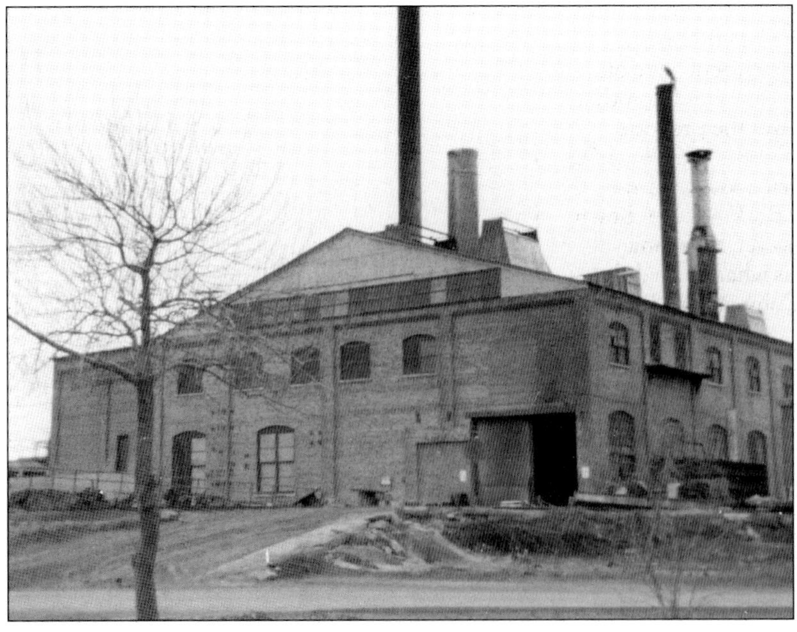

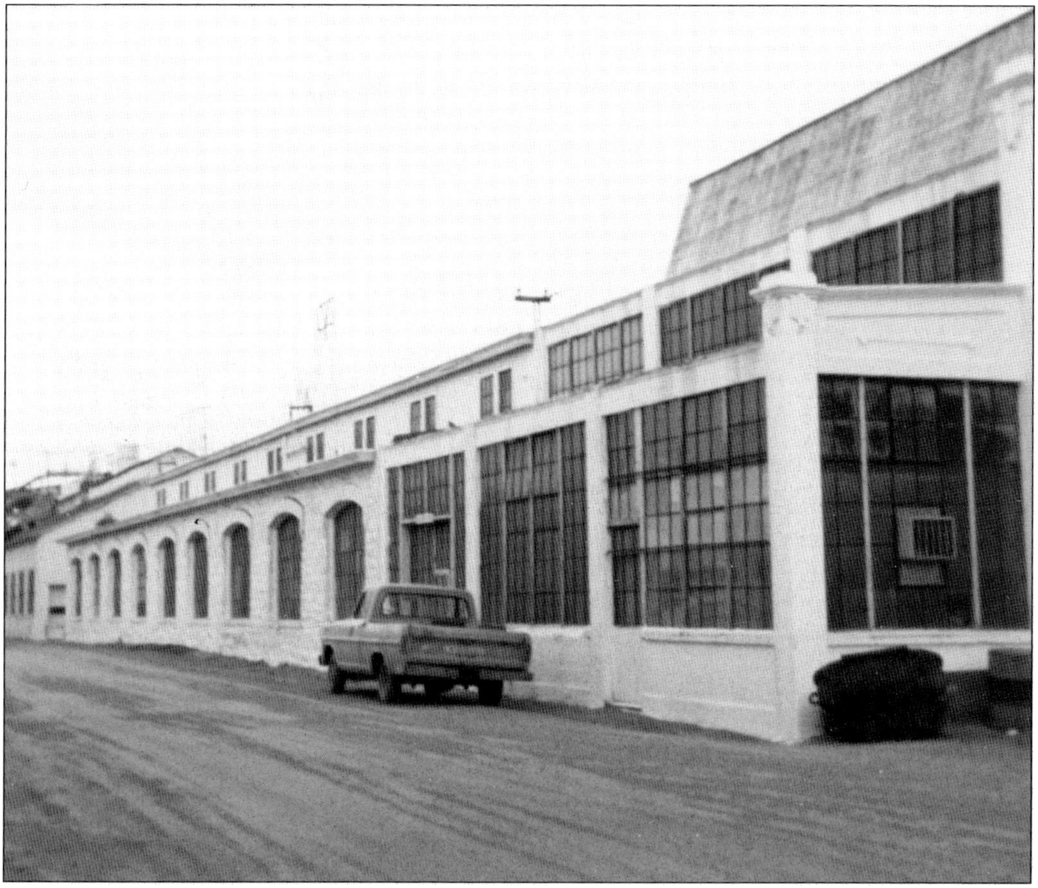

Health and safety became increasingly difficult to promise at the plant. The buildings as well as the equipment were old and outdated compared to other companies, and the processes were now different. The Machine Shop is shown above, and the flue and the Big Stack are shown at right. However, economic efficiency was what pushed the Anaconda Company to the edge. In November 1971, the Anaconda Company announced that it would require $23 million to bring the Great Falls Zinc Plant to the point it could remain functioning. In January 1972, the closing of the Great Falls Zinc Plant was announced. (Above, 1992.086.0002-65; right, 1992.086.0002-70.)

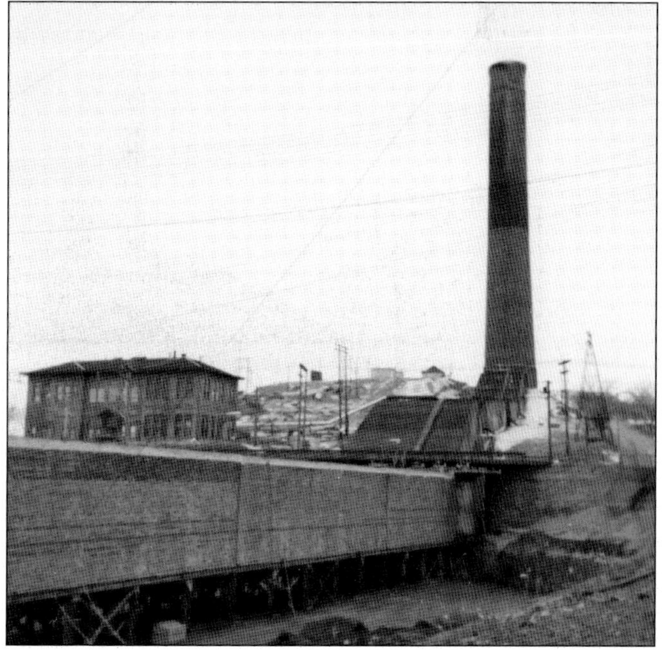

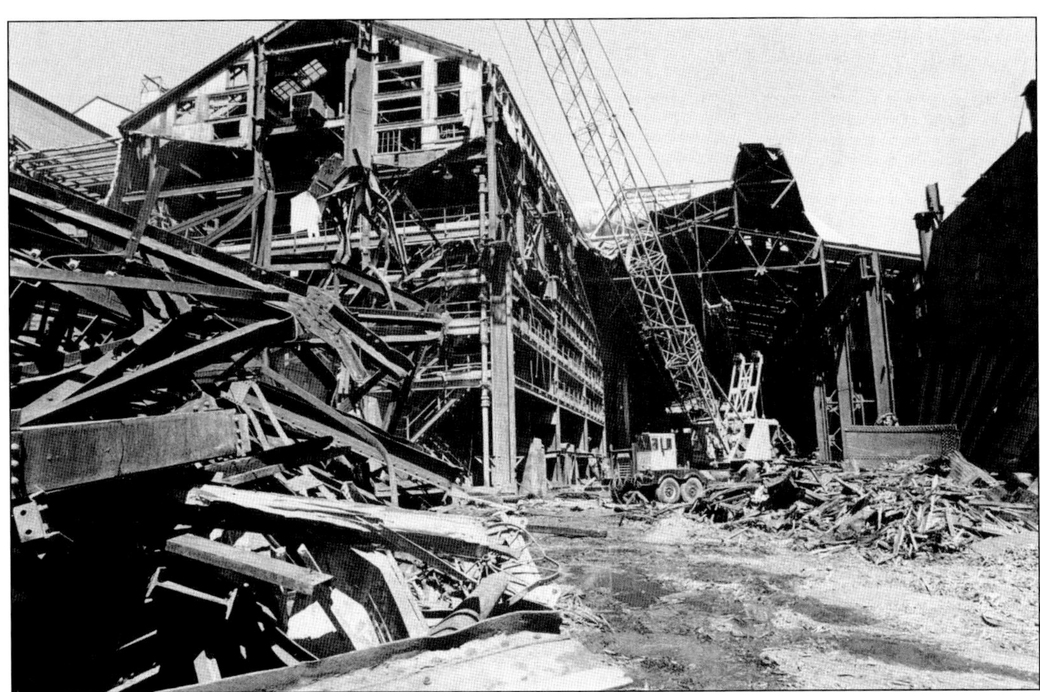

As if 600 workers losing their jobs were not enough, in May 1972, another 60 jobs were lost when the aluminum wire mill operations were also closed. Beginning in January 1973, the A&W Co. (Alaska Steel and Weissman and Sons) began demolition and salvage operations on the buildings of the zinc operations. These photographs, taken by Ray Ozman, show the tear down. The Big Stack would smoke no more but was left standing. In April 1973, John B.M. Place, the president of the Anaconda Company, paid the Great Falls Plant a visit. He told the *Great Falls Tribune* that "We expect to . . . expand production of the copper refinery in Great Falls." He also stated that "the copper refinery here is the company's most important in the nation." (Above, 2003.033.0017; below, 2003.033.0018.)

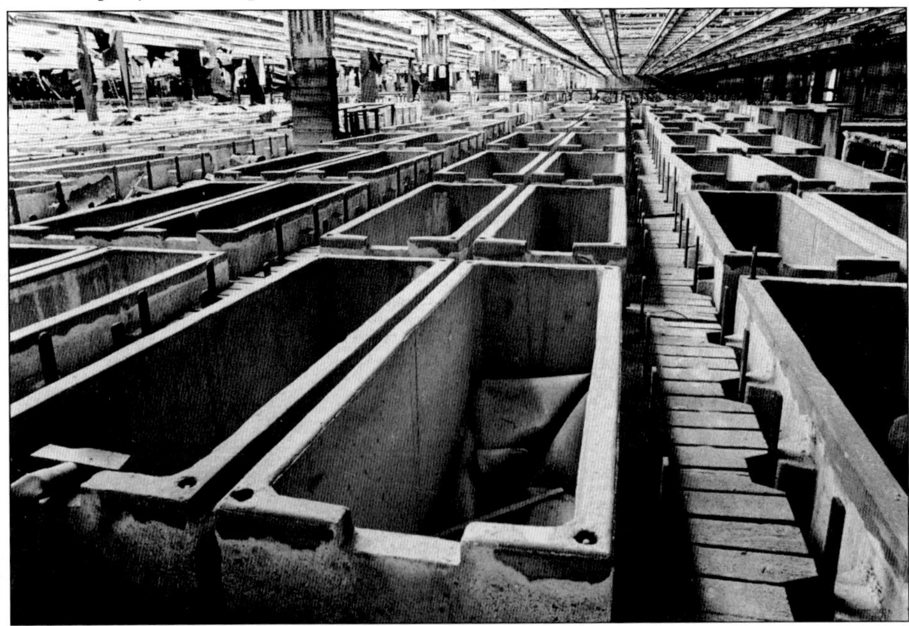

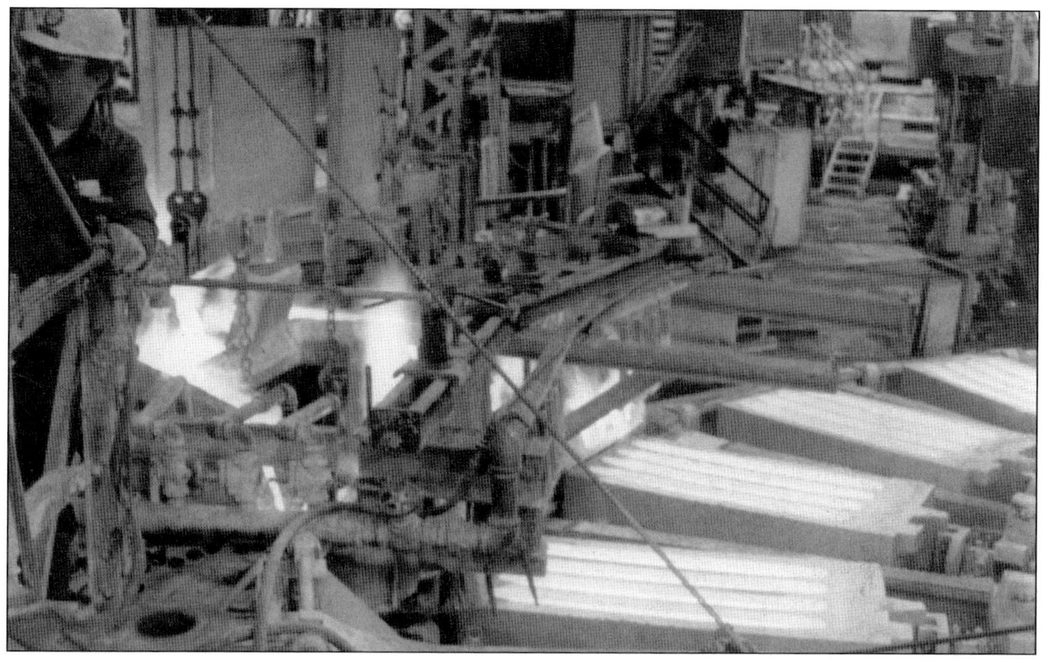

As the zinc buildings came down, a $7 million expansion was announced for the Electrolytic Copper Refinery in Great Falls. An extension of the older copper refinery building was enclosed so that construction work could continue through winter. These photographs show the bar molds being filled with molten copper. The project added a fourth circuit of electrolytic cells and a new starting sheet circuit. Also added were a new electrical substation and a drive-in entrance to the more spacious basement section on the northeast corner. (Above, 1982.017.0155; below, 1982.017.0157.)

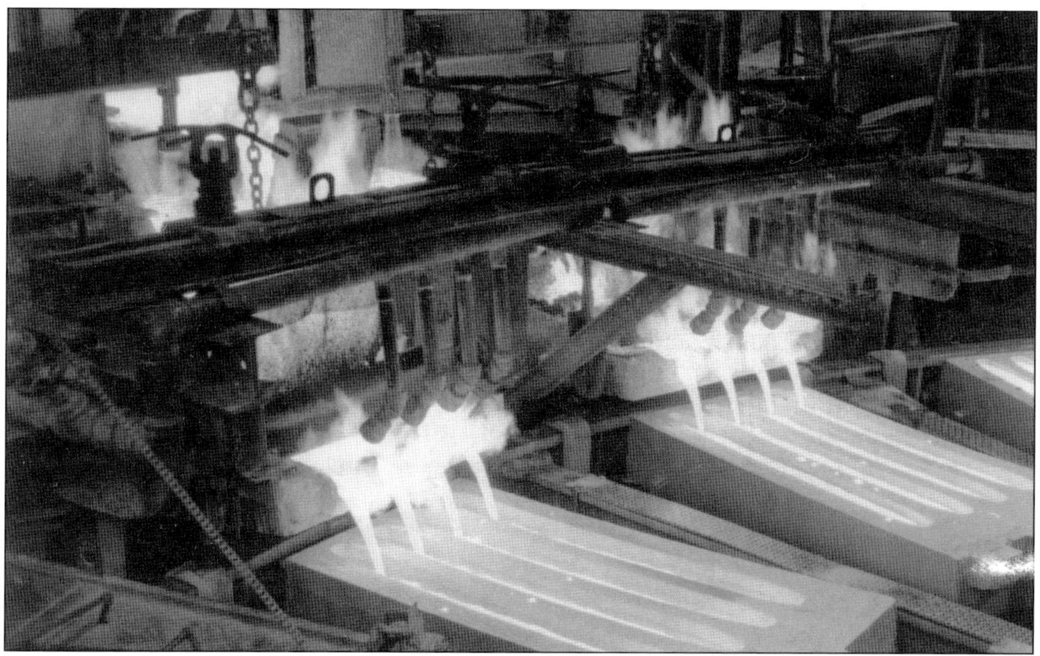

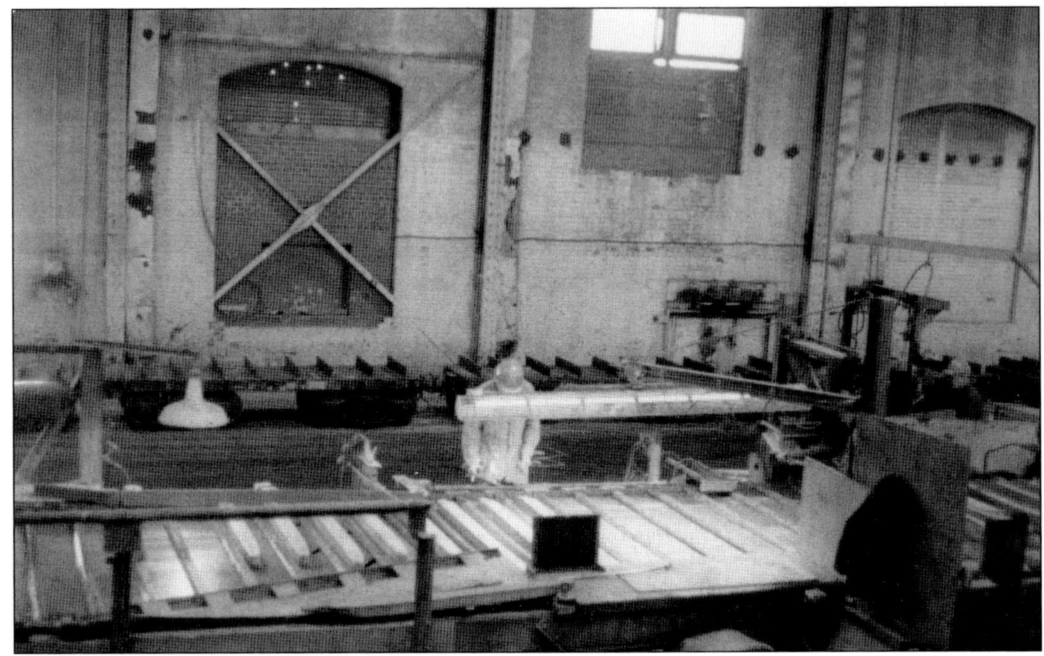

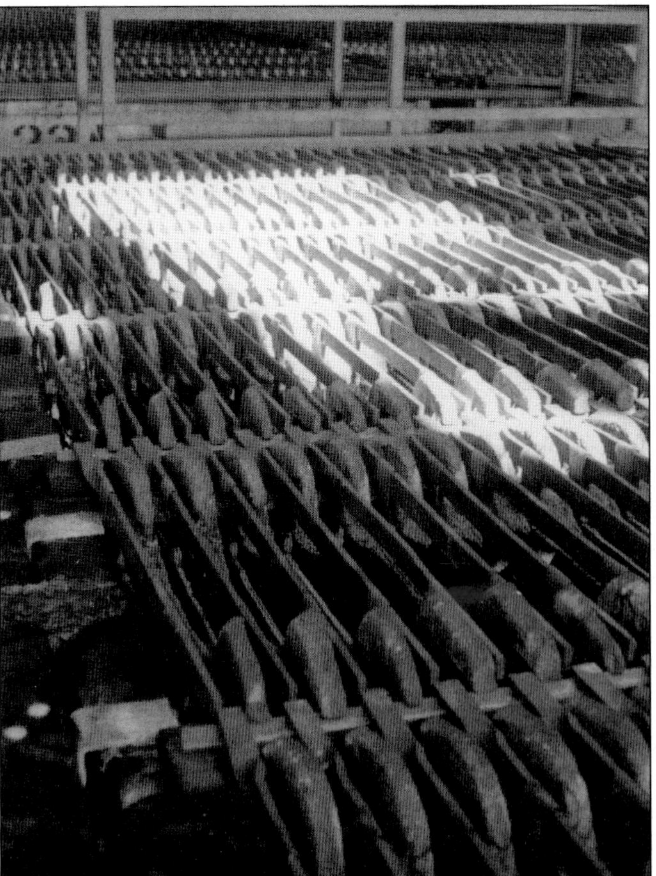

In the basement, concrete walls that supported the weight of the refining tanks were replaced with concrete pillars, which made maintenance significantly easier. Plastic-coated, lead-lined tanks were also installed to replace the old concrete ones. Production increased 38 percent, which equaled almost 12 million more pounds of copper. At left is a close-up of the copper electrolyte tanks filled with cathodes and anodes. The company forecasted hiring another 100 workers. Work seemed good at the Electrolytic Copper Refinery. Above, copper bars are being quality checked. (Above, 1982.017.0172; left, 1982.017.0163.)

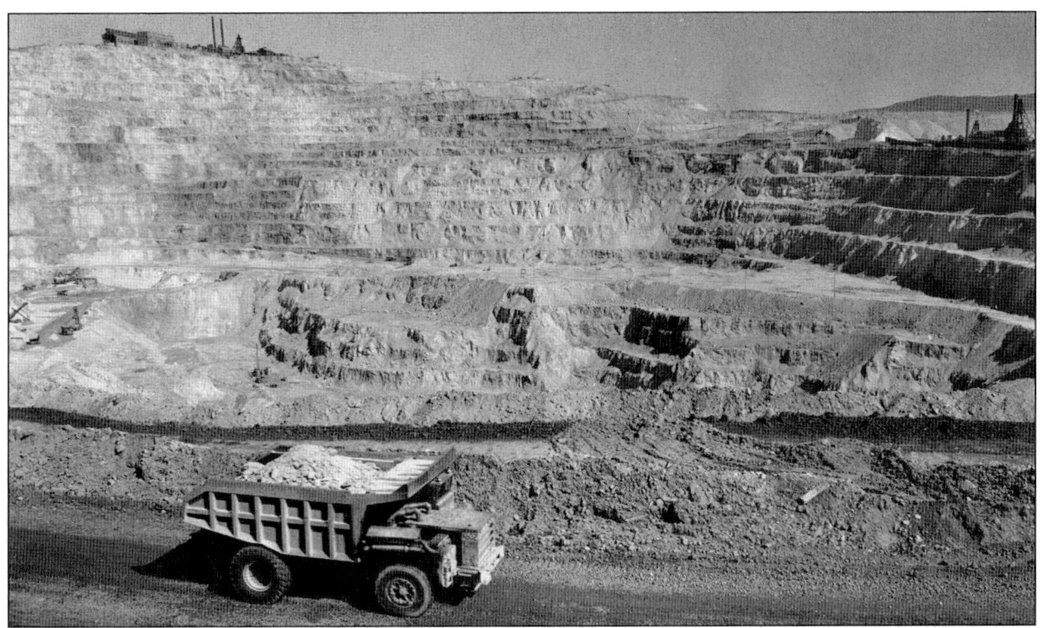

Supply of copper continued to fluctuate, causing layoffs and rehiring off and on through the 1970s. Above is a photograph of the Berkley Pit, the Anaconda Company mine in Butte, Montana, and below is the Washoe Smelter in Anaconda, Montana. The company golf course was leased to the City of Great Falls in February 1976. The Anaconda Company struggled due to the economy and attempted mergers with various companies, like Tenneco, until finally Atlantic Richfield (ARCO) purchased the company. For a moment, there was hope again, but it was not meant to be. (Above, 2003.033.0070; below, 2015.089.0007.)

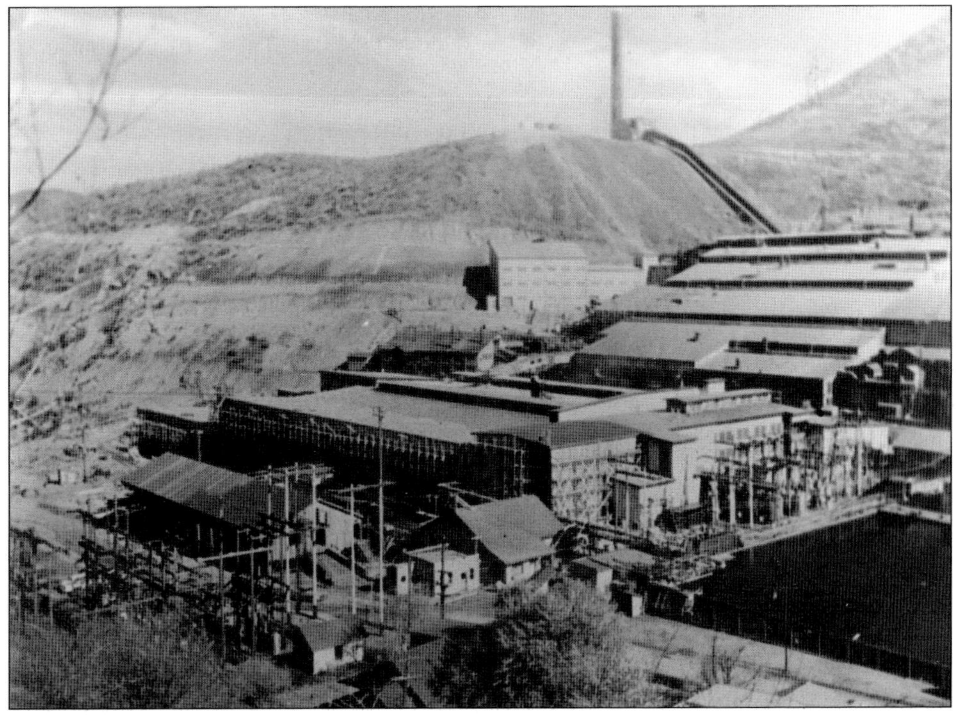

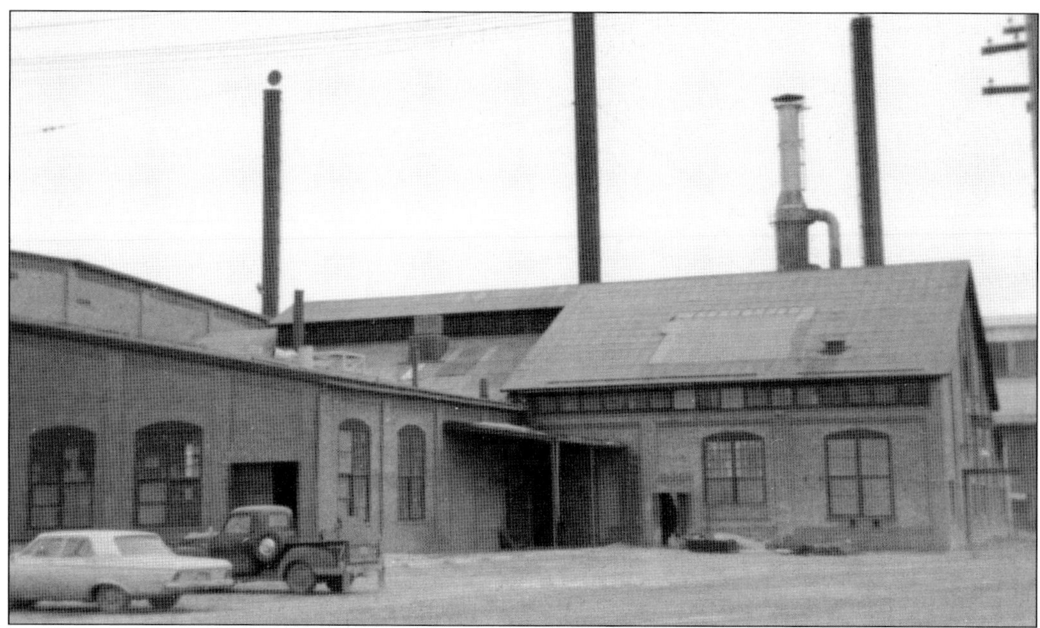

On September 29, 1980, Anaconda Copper, now a subsidiary of ARCO, announced the closing of the Washoe Smelter in Anaconda, Montana, and the plant in Great Falls. It was a loss of 1,500 jobs between the two towns, an unexpected blow to the state economy. ARCO stated that the closure was due to the high cost of complying with the federal air pollution standards that came about from the Clean Air Act in 1979. It would have taken almost $400 million to bring the Washoe Smelter in Anaconda, Montana, up to the new standards, and that was too costly. The Billet Casting Building is shown above, while the General Lab Building is shown below. (Above, 1992.086.0001-19; below, 1992.086.0001-51.)

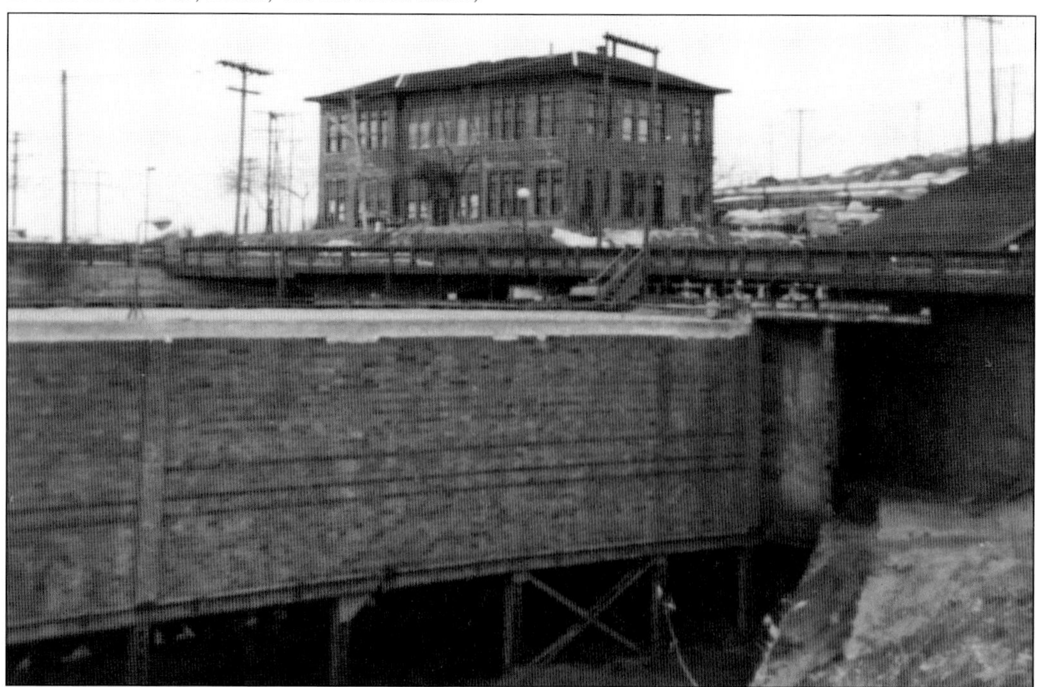

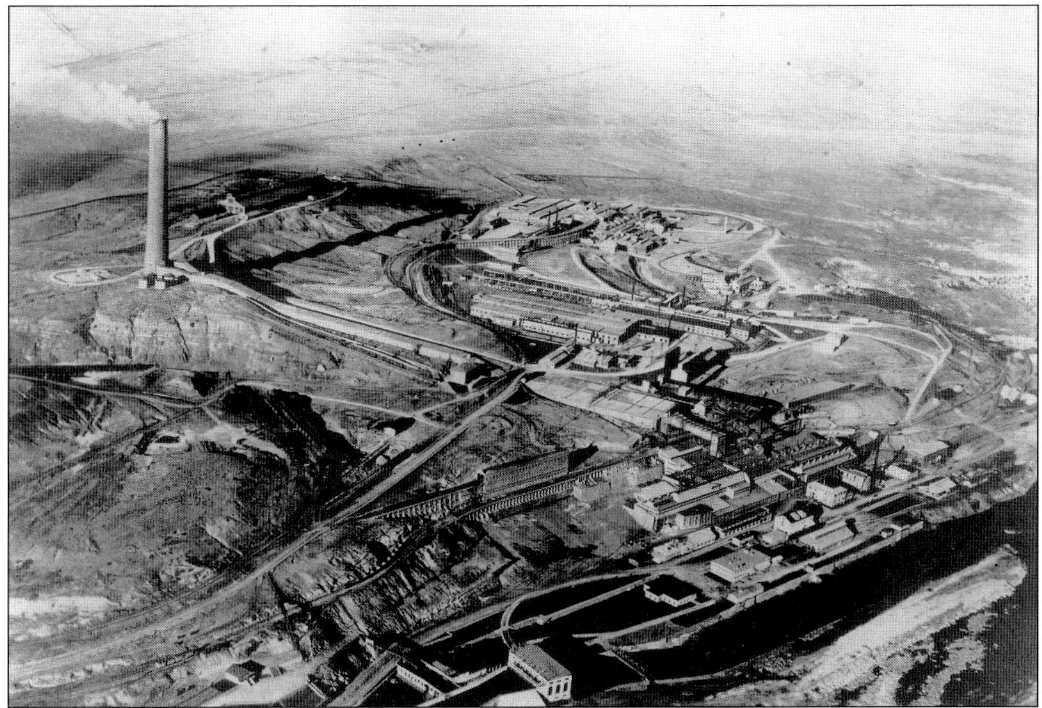

Though the plant had violated the air health standards 45 times in 1979, the EPA said there were mechanisms available through the Clean Air Act that would extend the deadline for compliance up to 1988, a provision available only to the copper industry. However, ARCO had made its decision, and there was no going back. The Great Falls Reduction Department Plant is shown from the air in the image above. A deal was made with the State of Montana and the City of Great Falls receiving $1.5 million from the Anaconda Company to stimulate new businesses and jobs. Below is the Employment Office Building. (Above, 1982.017.0035; below, 1992.086.0001-22.)

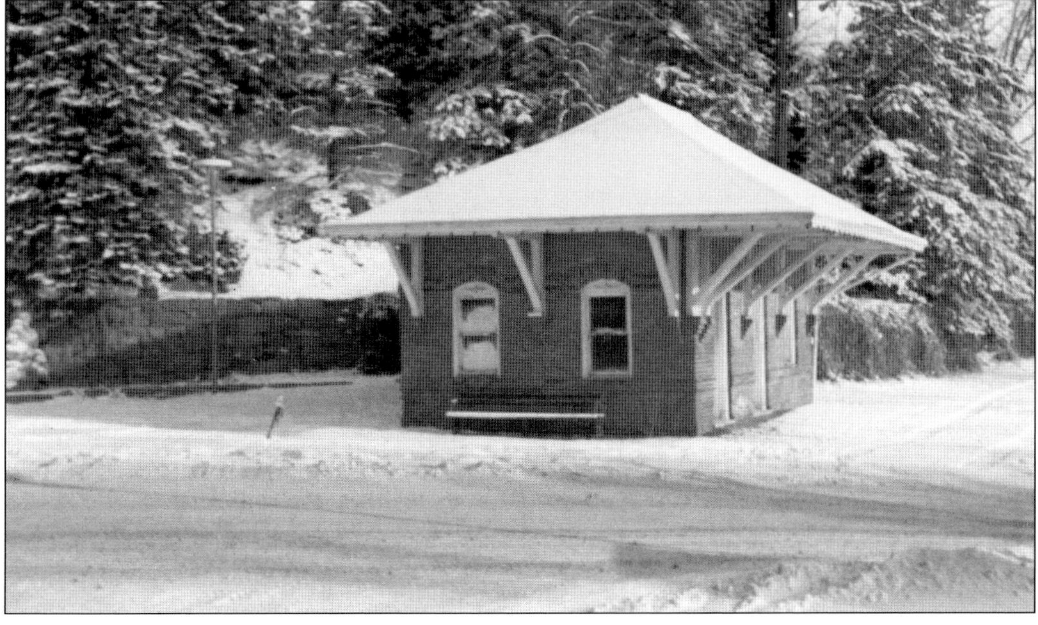

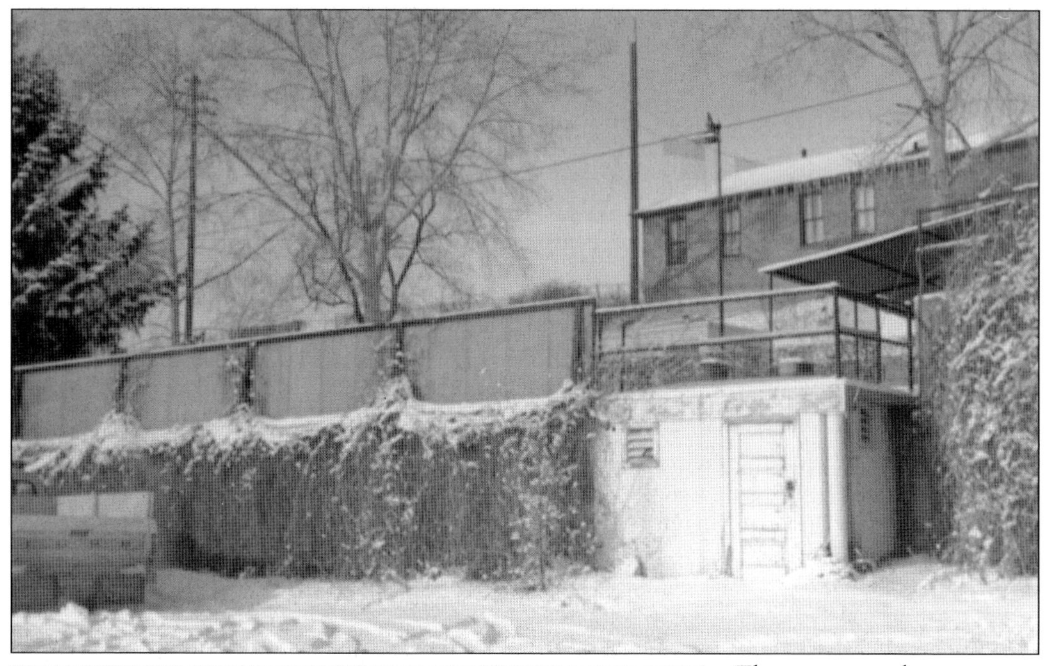

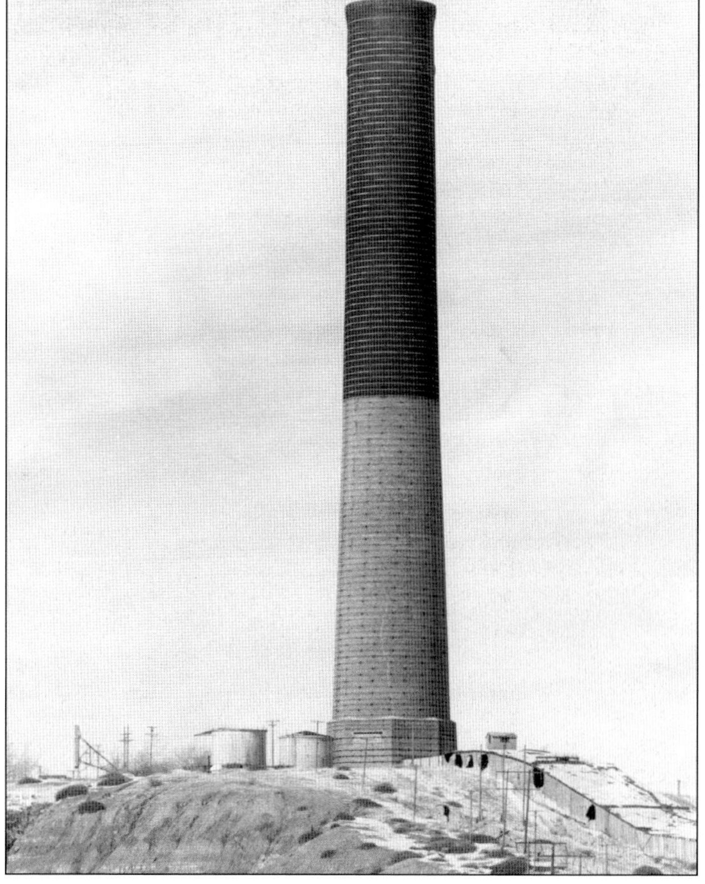

The payout made some people bitter, thinking it should have gone to the workers directly to help with job retraining and resettlement. Few accepted the reason given for the closure. The plant was dark, empty, and unused. Buildings were stripped bare and the grounds fenced off, as was the company pool, above. People figured that the buildings would be torn down eventually, but that was not what was chosen to go first. The Big Stack, as photographed by Wayne Arnst, is seen at left. (Above, 1992.086.0001-35; left, 2022.005A.0987.)

On November 19, 1981, it was announced that the Anaconda Company had contracted Pacific Hide & Fur to demolish the buildings and other structures of the closed plant, including the Big Stack. The alert was sounded, and dedicated citizens Bob Greenup and Greg Kecskes formed the Great Falls Stack Preservation and Historical Society. Petitions, phone calls, and donations—every effort was made to save the beloved landmark. The cover of the Big Stack booklet put out by the *Great Falls Tribune* is shown at right. Mutton Jeff even wrote a song, as featured below. (Right, from 2000.045.2; below, 1995.093.1.)

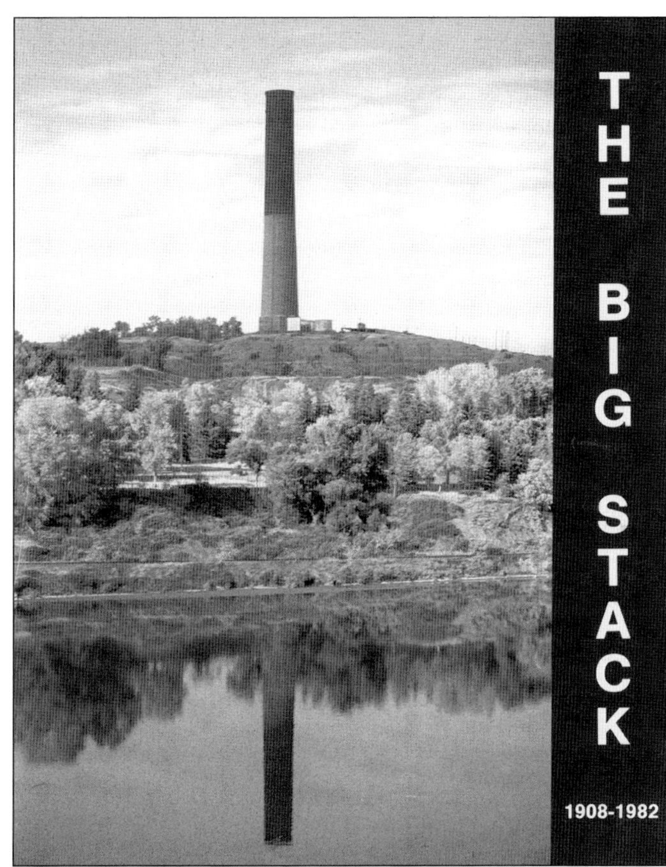

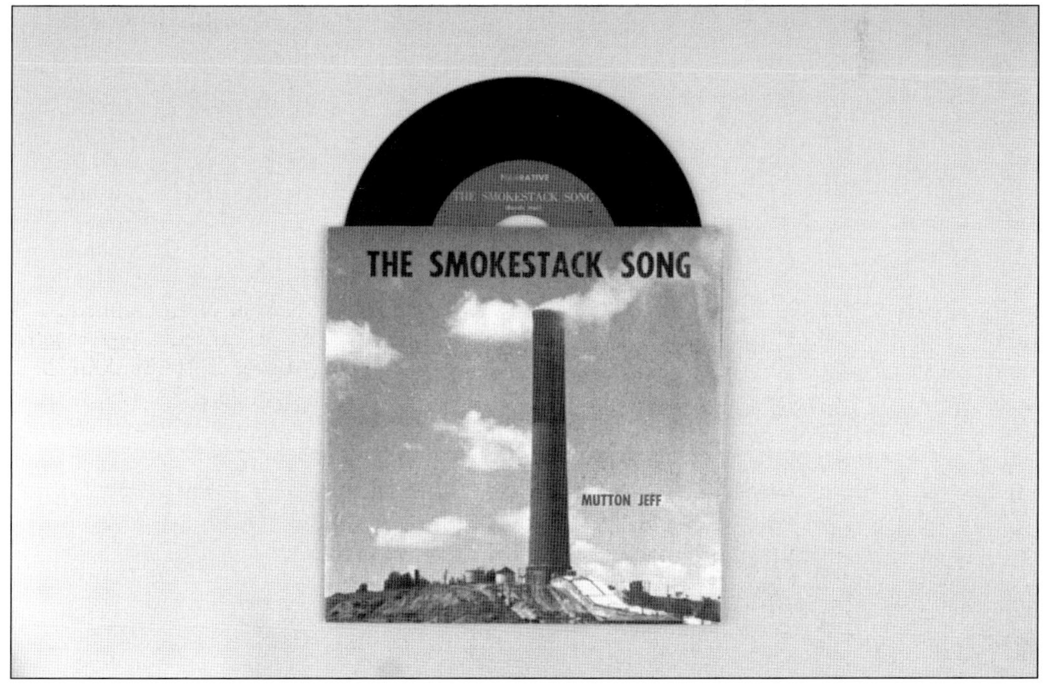

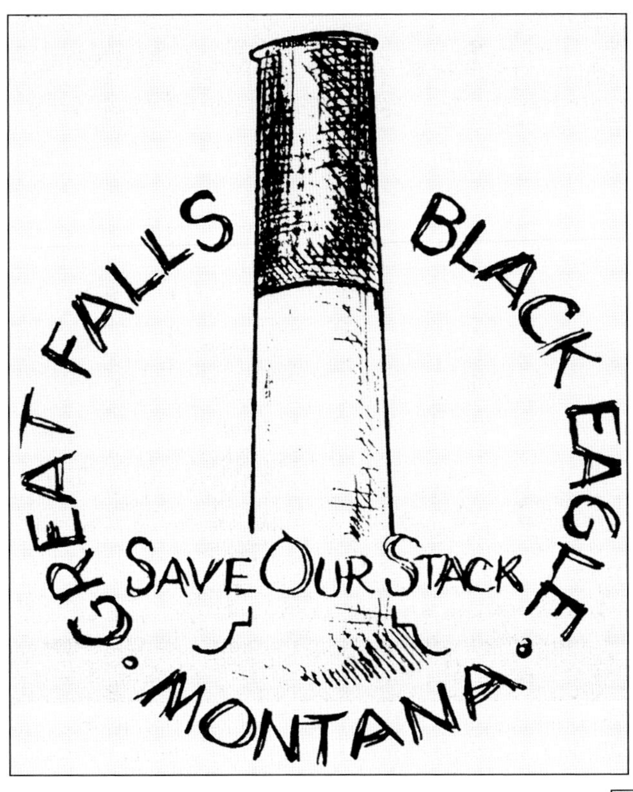

For a year, the call of SOS (Save Our Stack) went out to the world. Local cartoonist Mike Stuckslager drew the logo seen at left for the group and drew commentary for the local paper, seen below. It was a long-fought battle, with so many reasons to save or not to save the stack. Legal action was decided against, as it would just cause the group to go into debt. In one last-ditch attempt, a telegram was sent to Pres. Ronald Reagan begging for intervention, but no help came. The Big Stack was declared too dangerous to stay standing. The date for demolition was set for September 18, 1982. (Both, from 2000.045.2.)

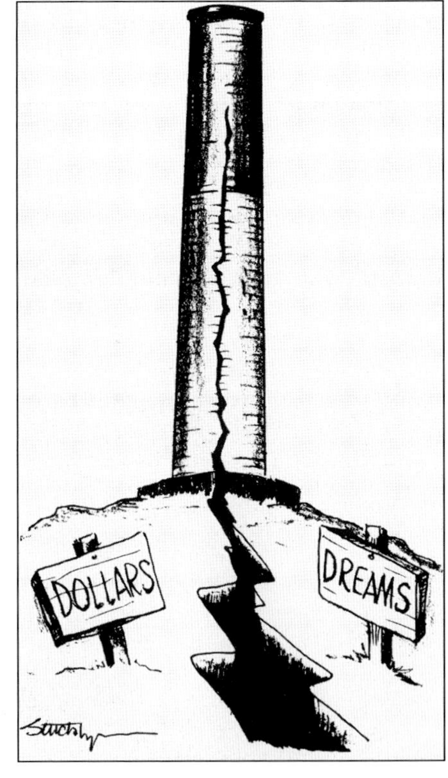

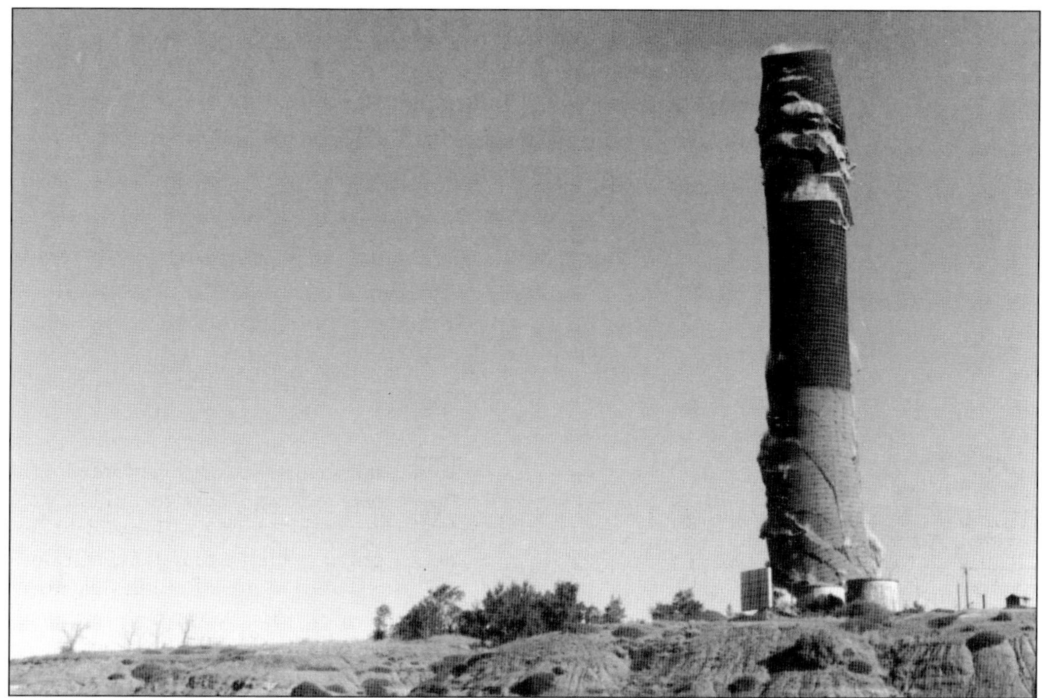

In the week leading up to the demolition, preparations were made by the demolition team. Plans were made just in case there was low cloud cover. The Federal Aviation Administration announced a three-mile-radius restriction around the stack. Early on the morning of demolition, several thousand people in the area searched for a spot to watch the Big Stack fall. The entire sheriff's department was called upon to be security around the Anaconda Company grounds and direct traffic in Black Eagle. (Above, 2014.073.0001; below, 2014.07.0002.)

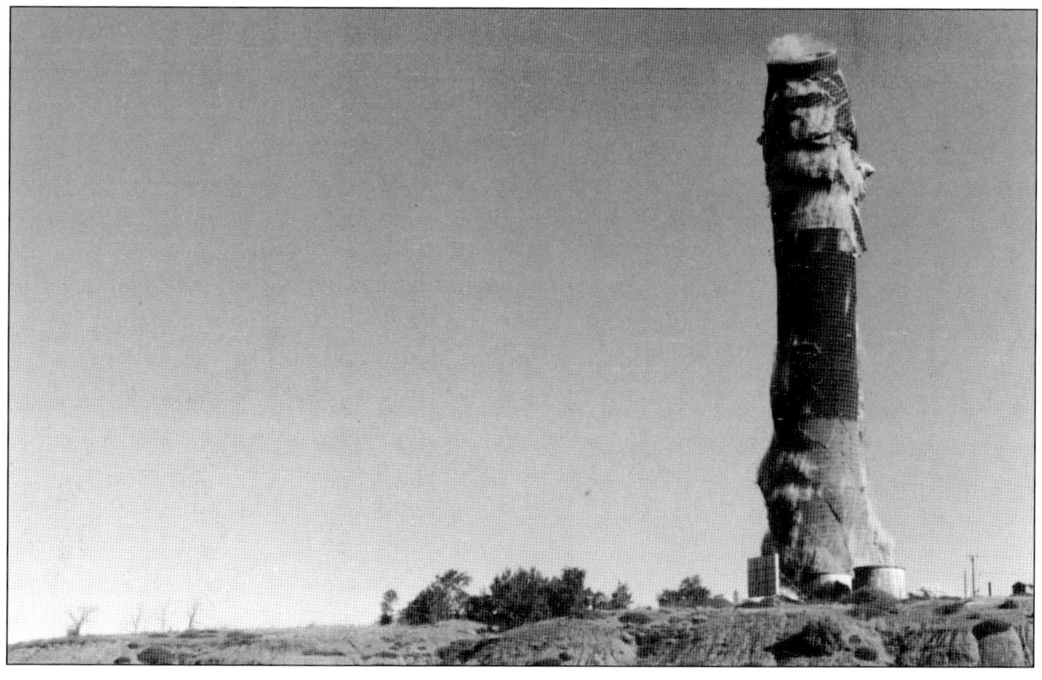

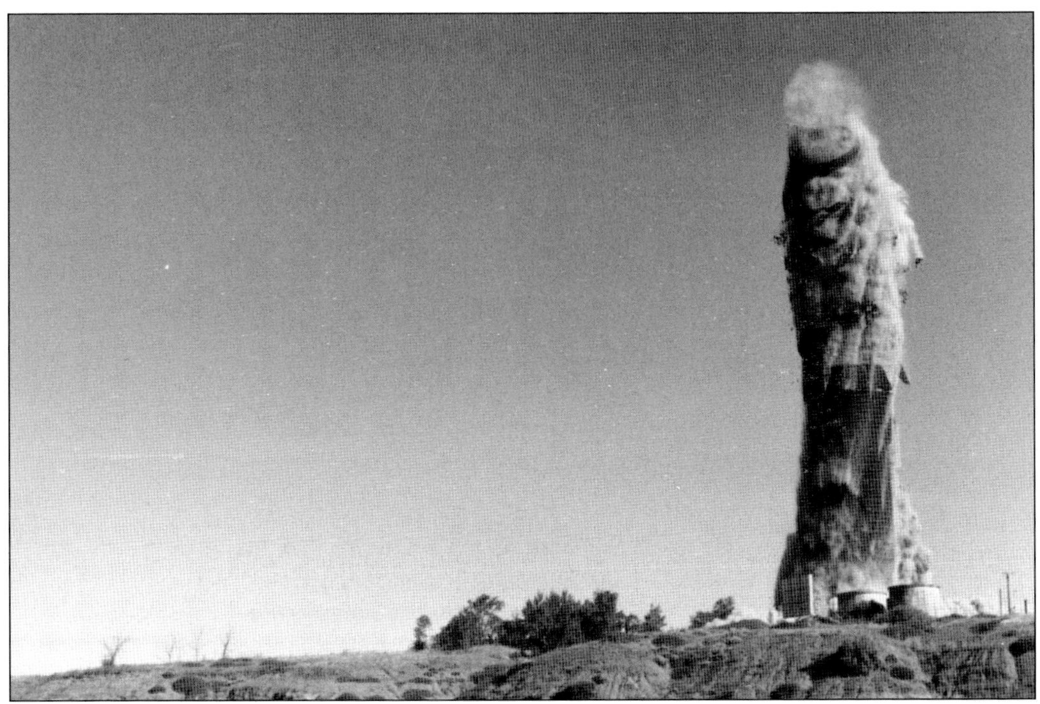

At 10:00 a.m. that Saturday, the whole community was waiting. Six hundred pounds of explosives had been placed to bring the Big Stack down. There was a thrill of excitement for some, but for others, it was liking watching a death happen. The Simmons Oil Refinery, just upriver from Black Eagle Dam, sounded its siren at 10:05 a.m. Silence filled the air, and five minutes later, the dynamite went off. (Above, 2014.073.0004; below, 2014.073.0005.)

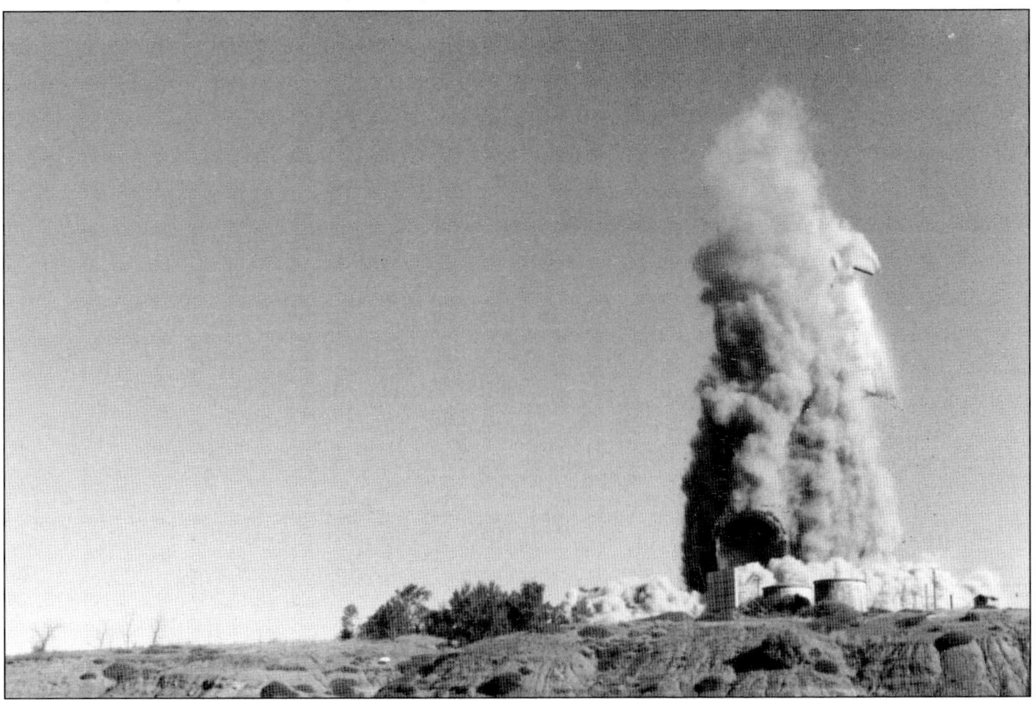

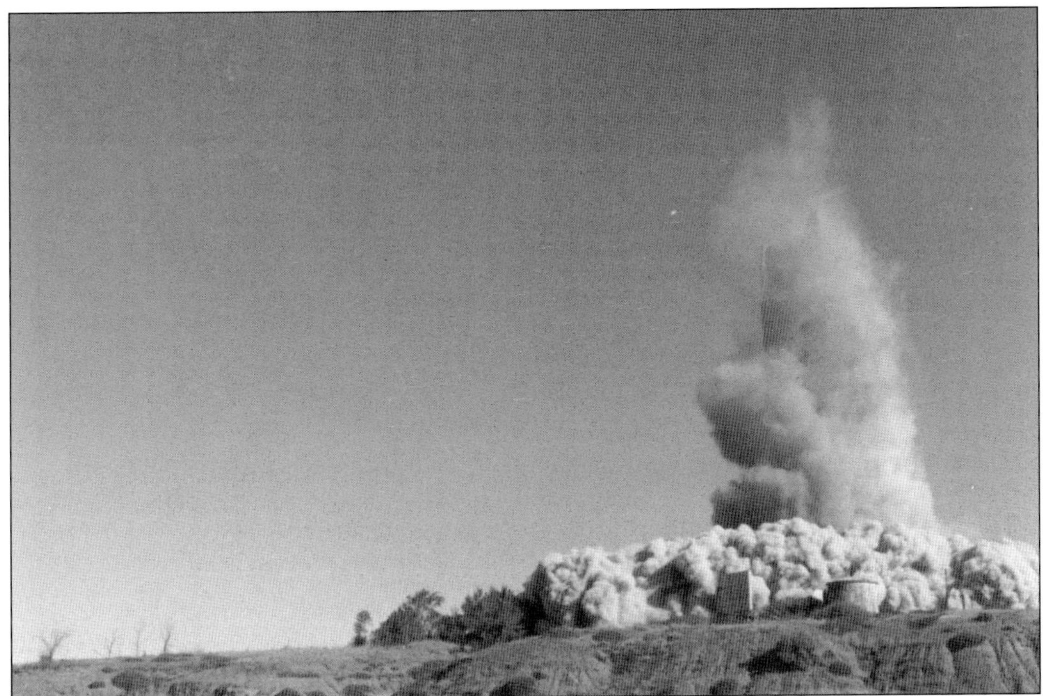

The blast was spectacular; the noise from the explosion rippled outward, and the stack sheered in half. The bricks and mortar fell, and dust rose from the ground. When the dust settled, a slice of the stack still stood on the skyline. Many of the onlookers cheered, loving the Big Stack's defiance. For many, it was confirmation of the stack's hardiness, proving it was not a danger to public safety like it had been portrayed. (Above, 2014.073.0008; below, 2014.073.0009.)

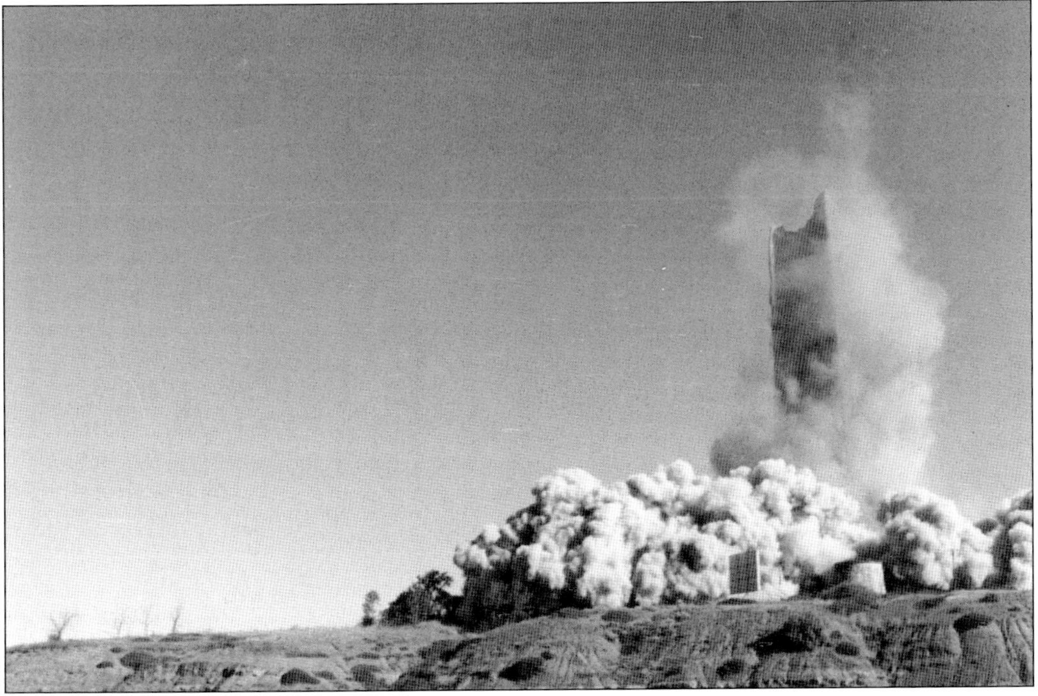

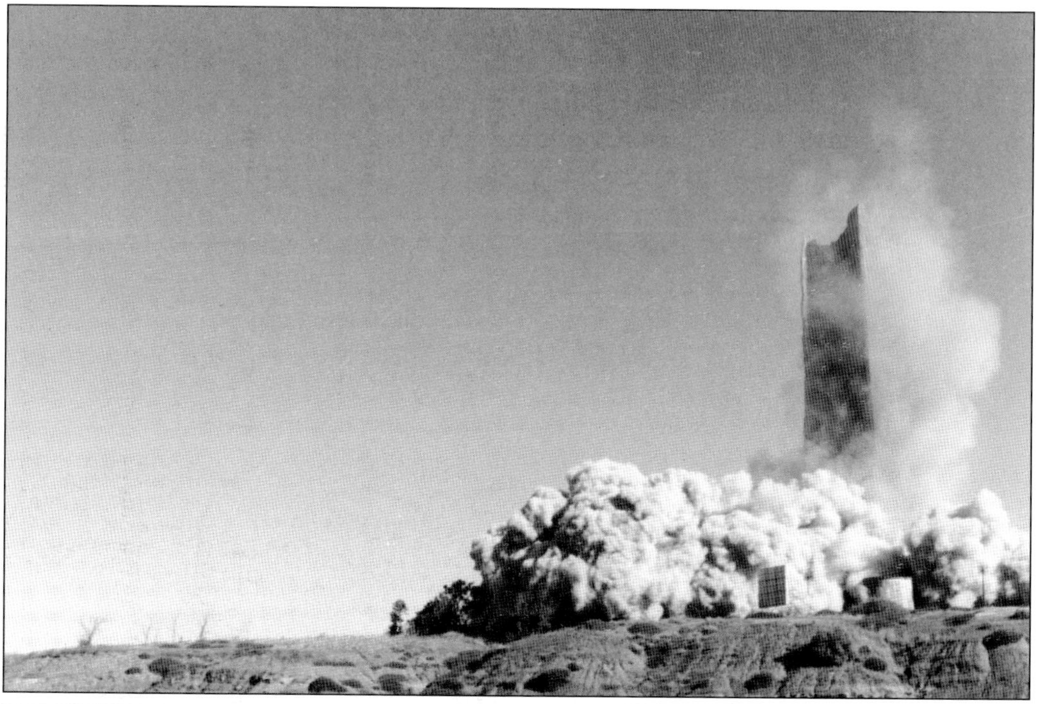

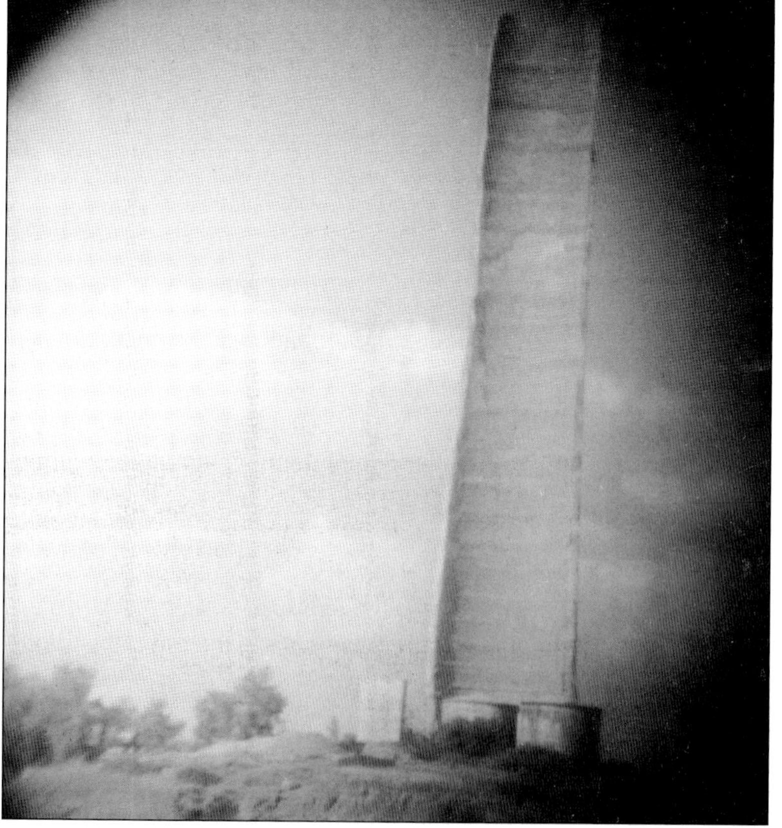

More explosives were brought in from Butte, Montana. The additional explosives arrived mid-afternoon, and 400 pounds were placed into the bored holes at the base. At about 5:00 p.m., the Simmons Refinery siren went off again. The splinter came down in seconds, an anticlimactic end to a disheartening event. (Above, 2014.073.0011; left, 1983.043.0012.)

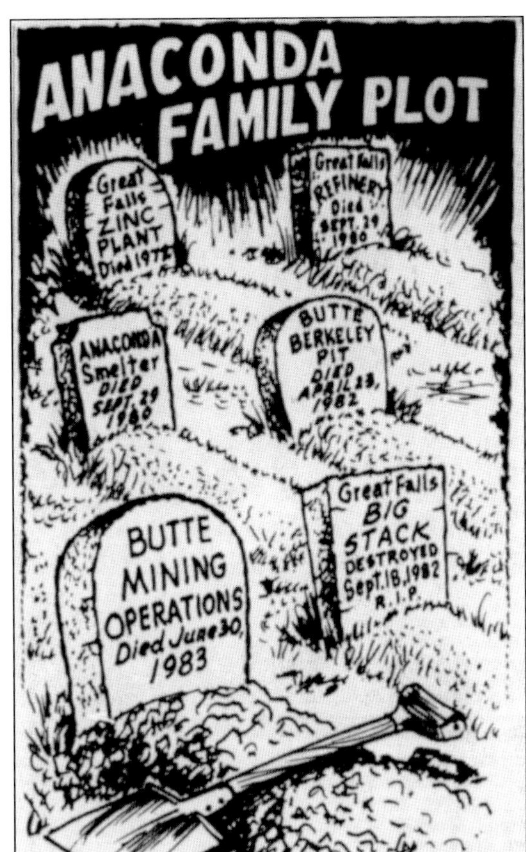

The icon of the city was gone. The bars that day were full, people reminiscing, some crying. It had always been there, like a light watching over travelers. Pacific Hide & Fur sold about 200 bricks from the demolished smokestack, with profits ironically going to the Great Falls Centennial Committee. Below is a 7-by-5.4-by-5.5-inch stack brick with plaque. (Right, *Great Falls Tribune*, July 1, 1983; below, 1982.064.0015.)

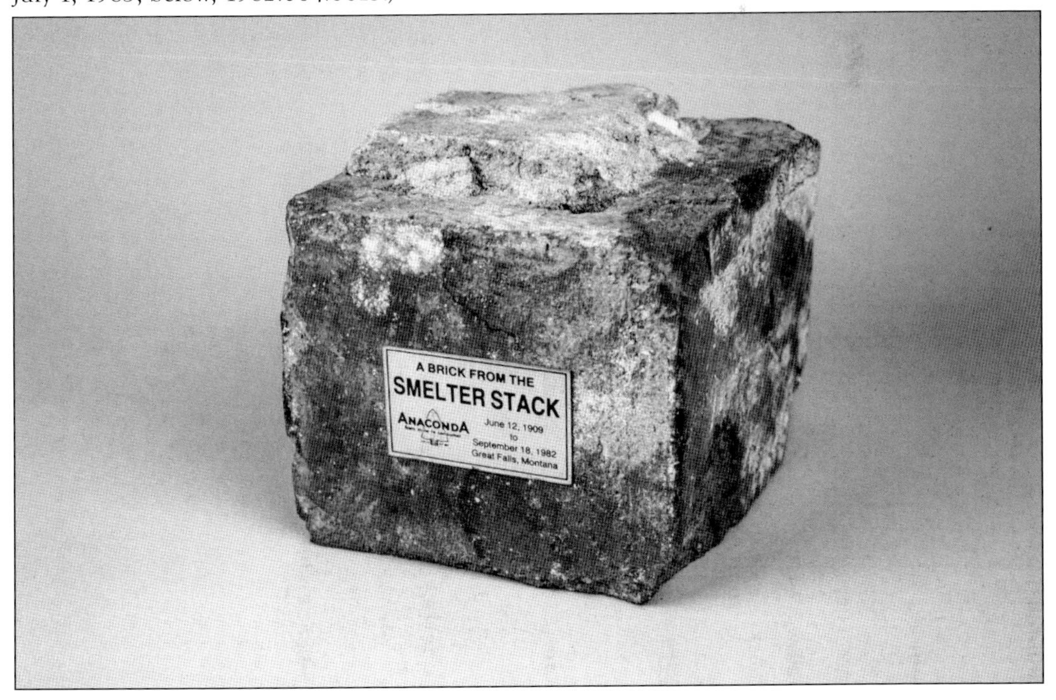

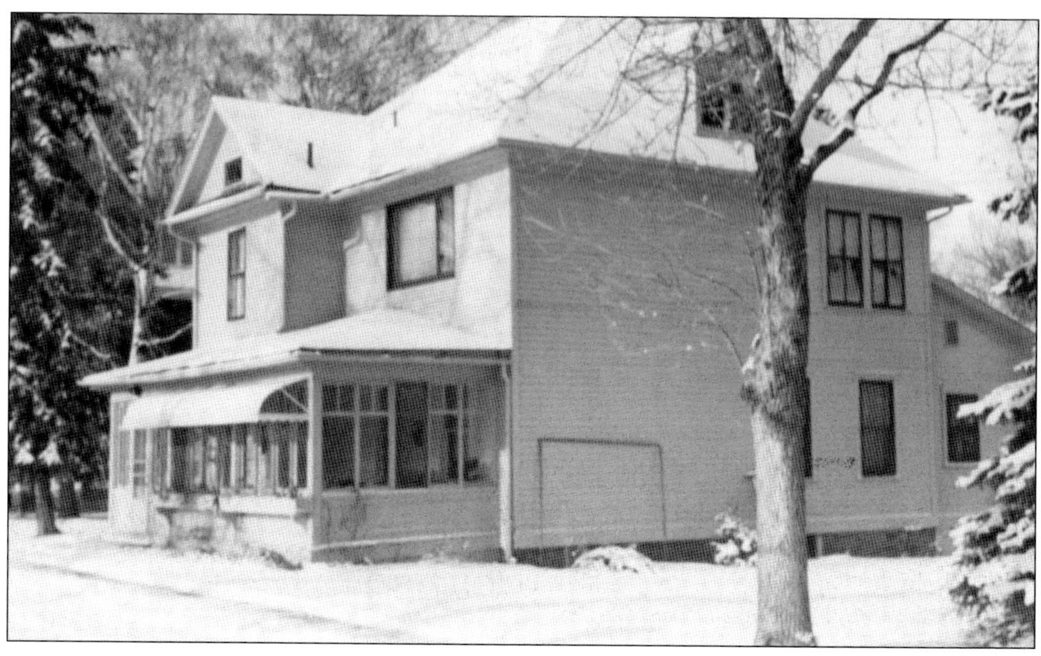

With the Big Stack dealt with, Pacific Hide & Fur moved on to the plant buildings. The company houses were offered a chance at redemption and were put up for sale, but they had to be removed from the property. Of the 15 homes available, nine were purchased. Then came the fun part of moving them to their new destinations. (Above, 1992.086.0002-7; below, 1992.086.0002-19.)

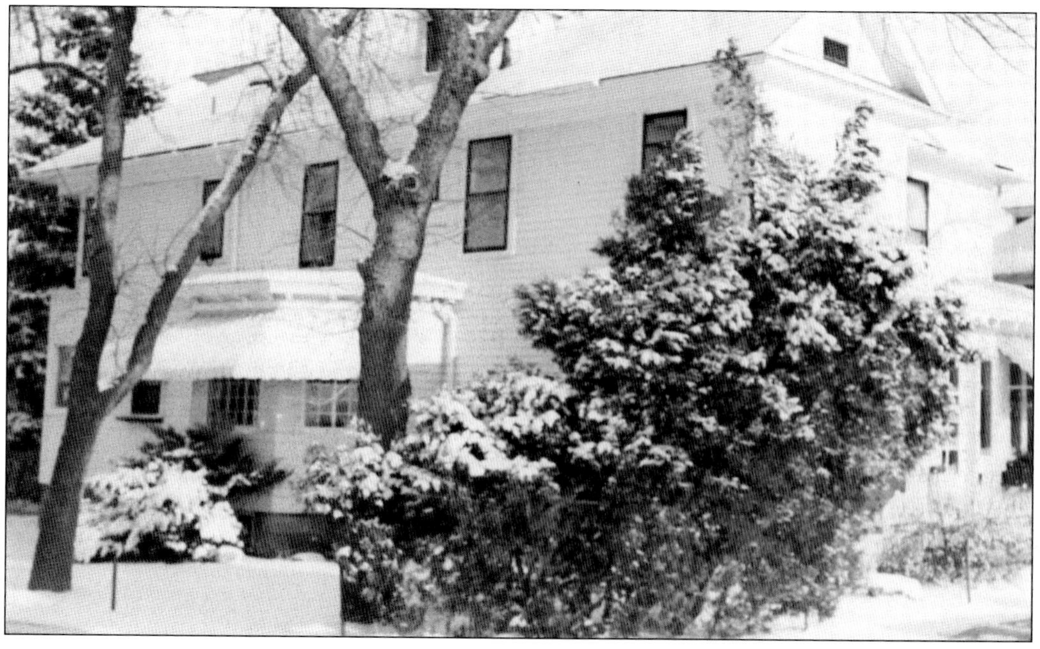

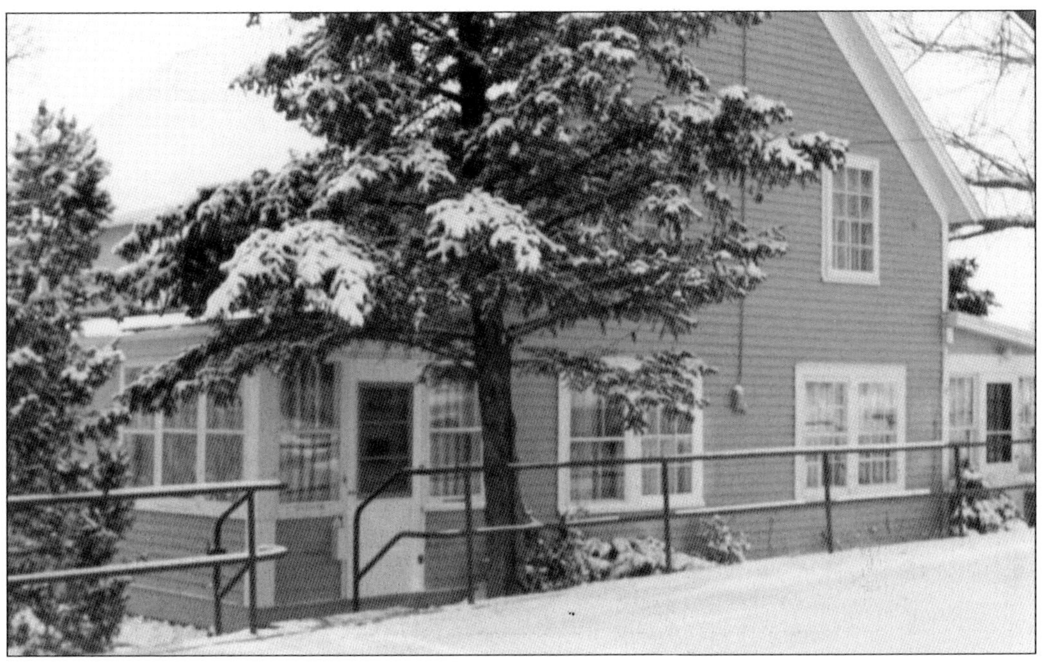

Three of the homes were purchased by Neighborhood Housing Services and placed on three adjoining lots on the south side of Great Falls in the hopes of revitalizing the area. One was purchased for a ranch east of Dutton, Montana; another went out just east of Fairfield, Montana; and another out went to a ranch northwest of Carter, Montana. The last three were moved in September 1983 to the lower north side of Great Falls; Vaughn, Montana; and Manchester, Montana. (Above, 1992.086.0002-2; below, 1992.086.0002-12.)

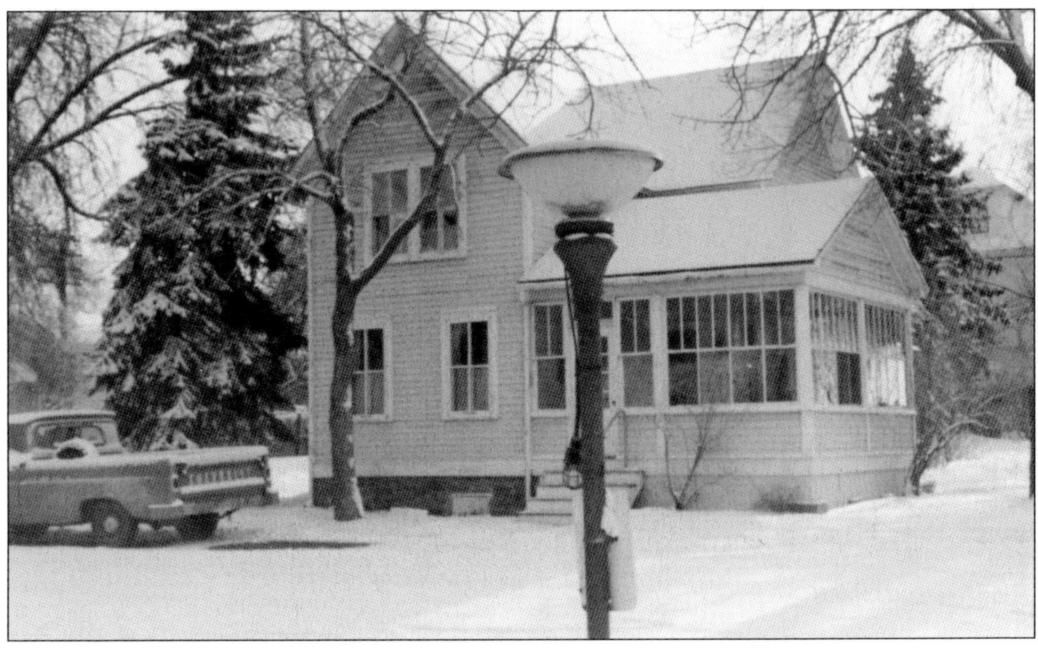

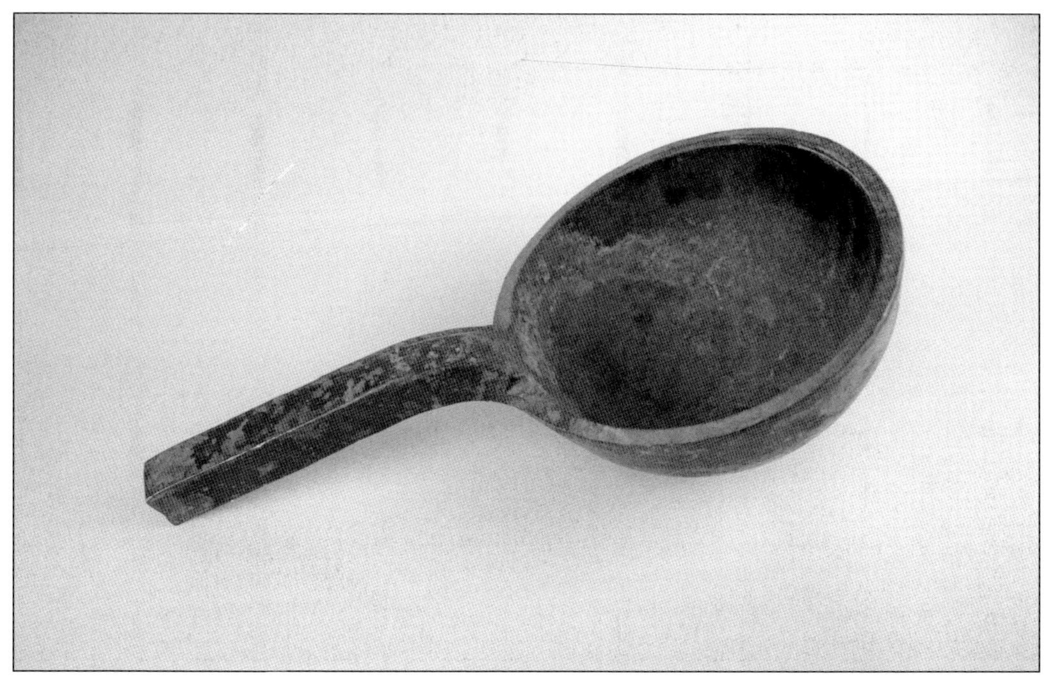

Smelter Hill was wiped away over a period of five years. Buildings were either taken down or demolished. Employees of both the plant and the salvage company found and collected items and records, knowing they were important. All the items featured on this and the following pages are housed at in Great Falls, Montana, thanks to the public's efforts. Above is a molten metal ladle, measuring 3.5 by 14.5 by 7.75 inches, and below is slag from the copper refinery, measuring 1.75 by 4 by 6.5 inches. (Above, 1982.027.24; below, 2020.036.1)

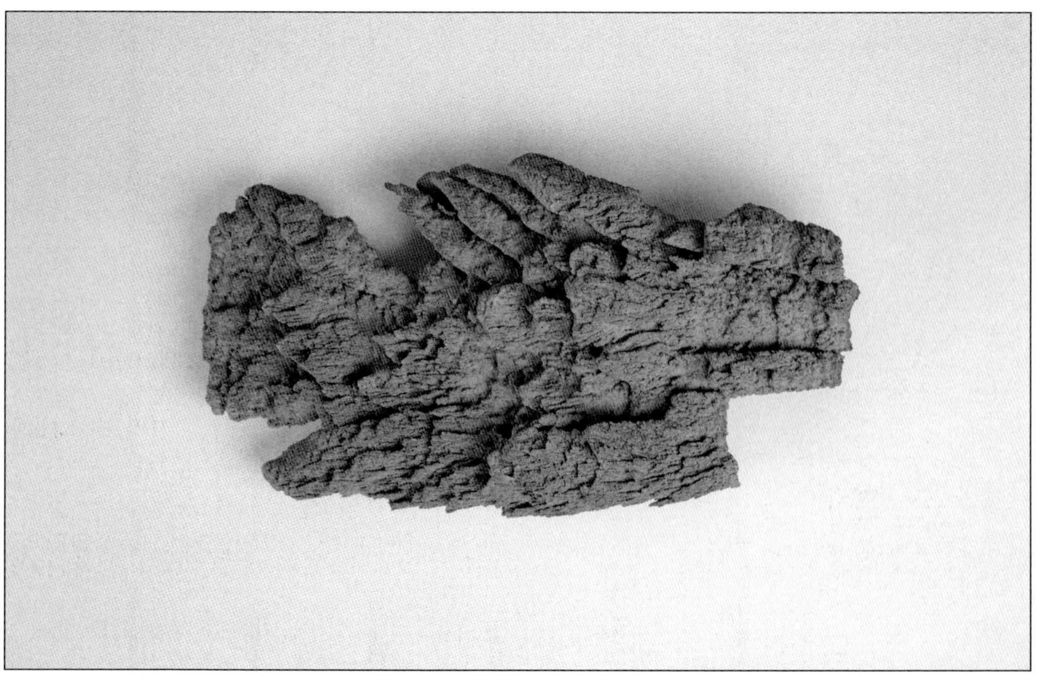

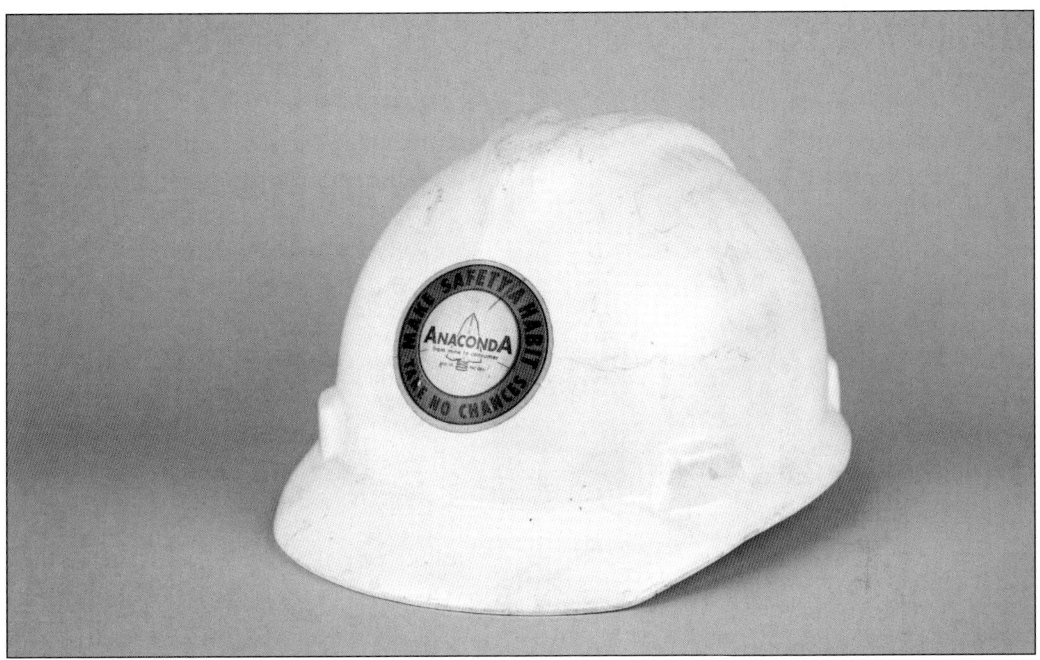

The Great Falls Reduction Department had been a huge part of the lives of not only Great Falls and Black Eagle but Montana as a whole. Remnants still linger in people's homes and businesses as well as their hearts. Many still tell stories of their time working at the plant to their children and grandchildren. Here are an Anaconda Company hard hat, measuring 6 by 8.5 by 11 inches, and a filter press cloth reused as a drop cloth for painting, measuring 30 by 58 inches. (Above, 1982.059.6; below, 2011.047.1.)

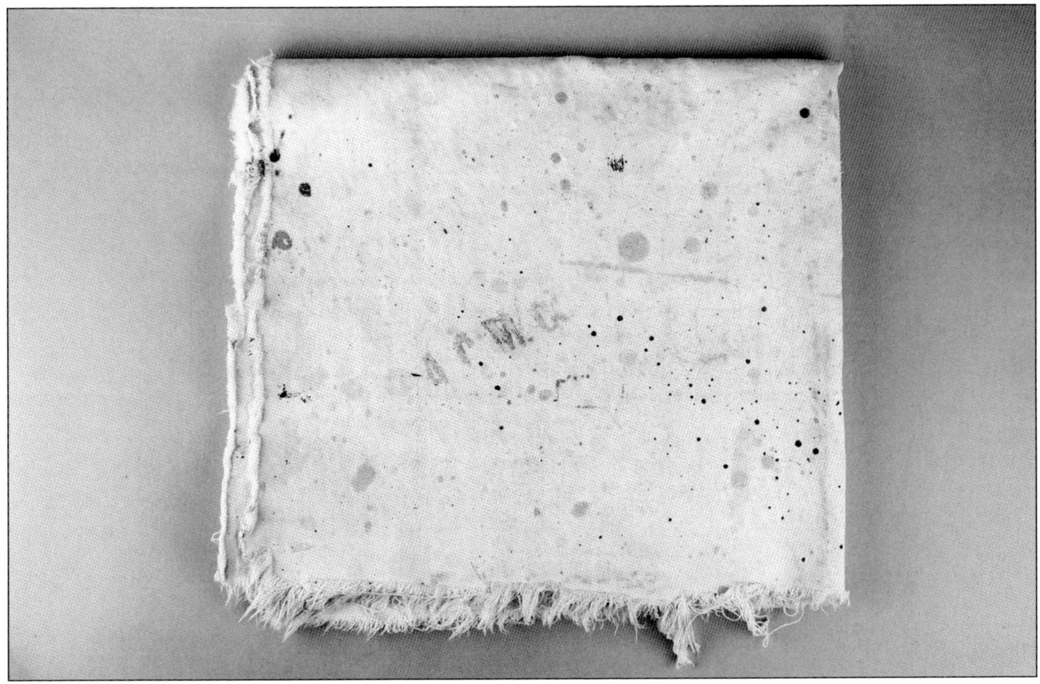

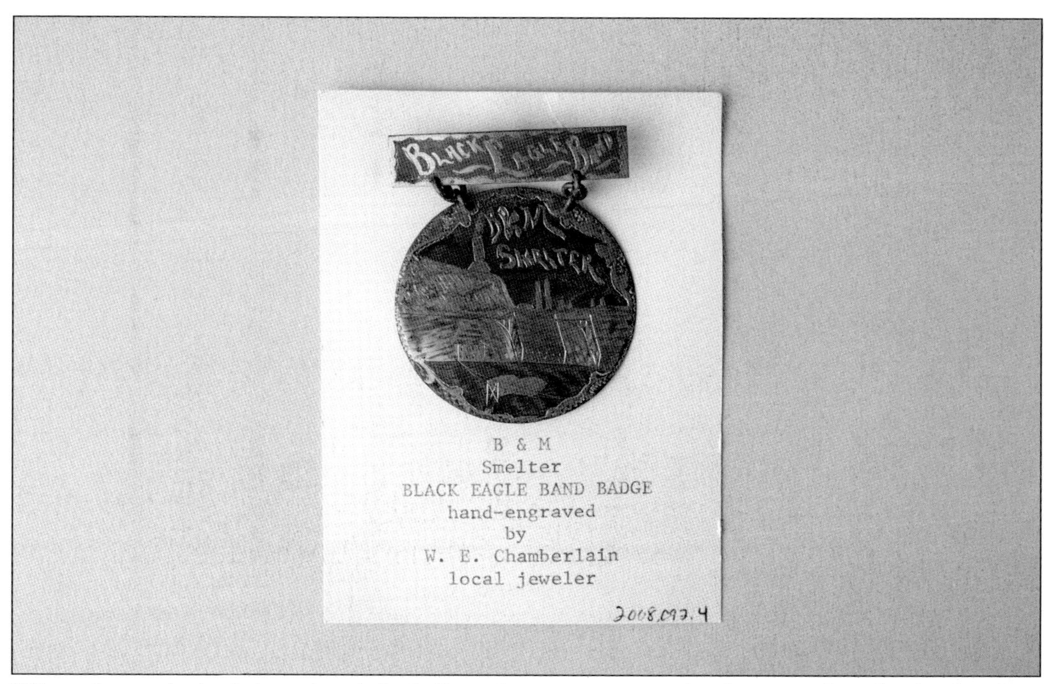

When the smoke had cleared and the hill was bare, it was time for the grounds to heal. The plant was declared a Superfund site, and the Atlantic Richland Company made efforts to work with the Montana Department of Environmental Quality to decide what remediation was needed. In 2000, evaluation by the EPA looked at the community of Black Eagle first to see if there was contamination of properties and whether it created a health risk or an ecological one. Above is a custom-made Black Eagle Band badge featuring the Boston and Montana Smelter, measuring 1.75 by 2.25 inches, and below is an aluminum bar, measuring 1.25 by 10.25 inches. (Above, 2008.092.4; below, 1983.102.4.)

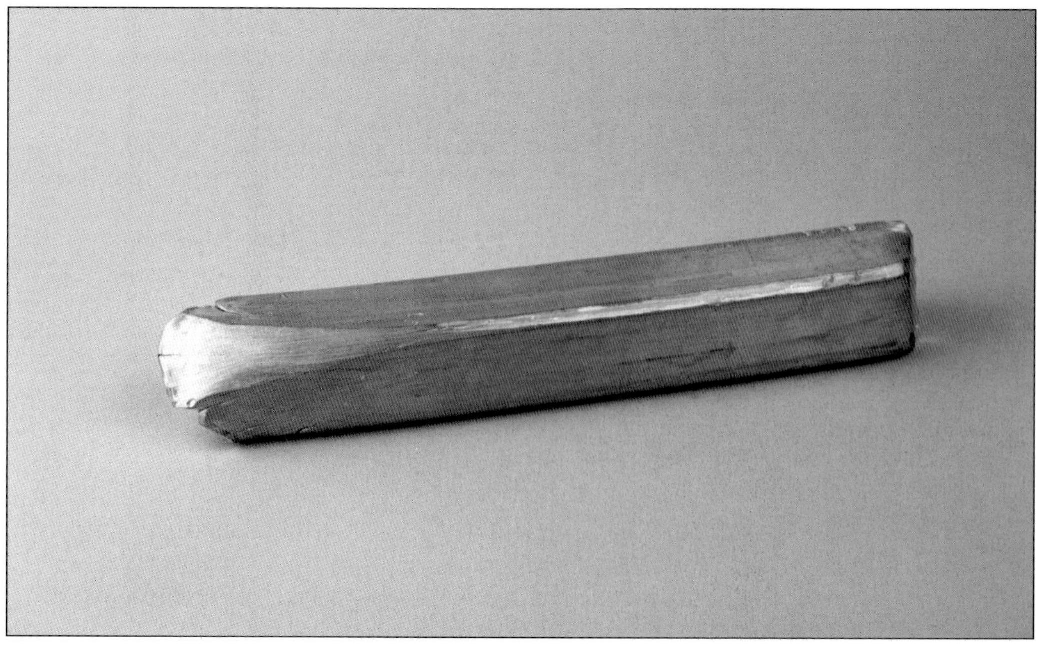

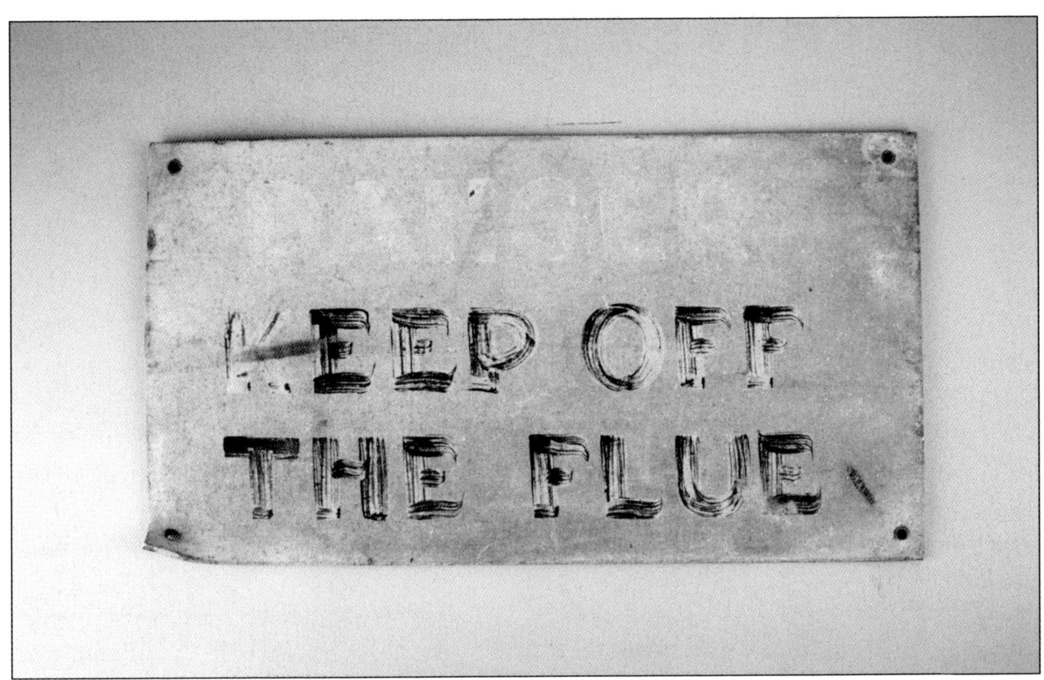

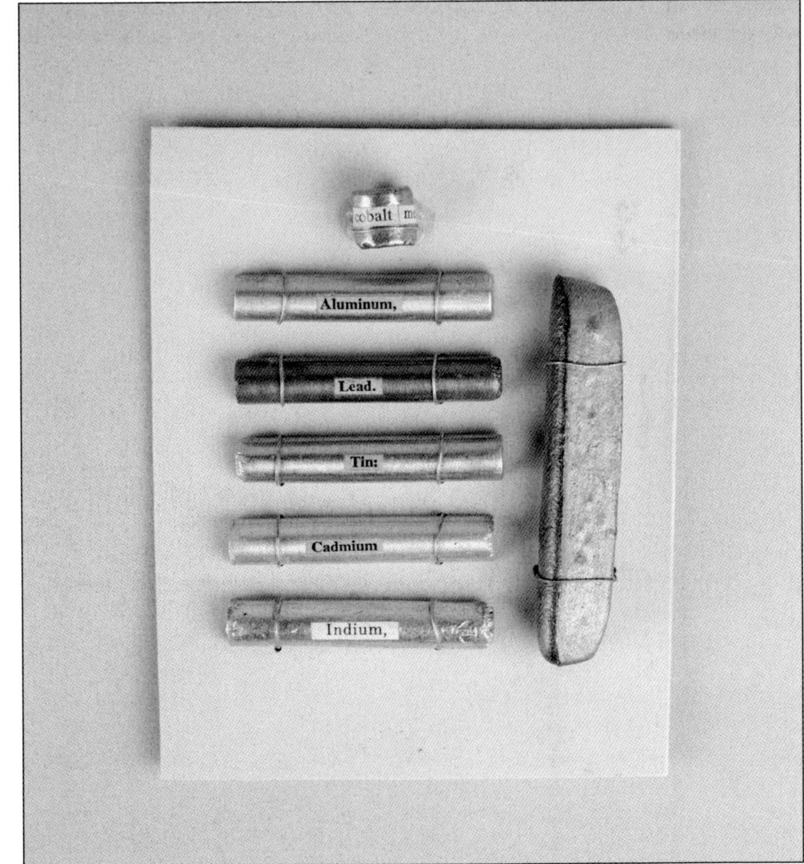

In 2011, after testing the sediment, surface water, groundwater, and air samples of 500 individual properties, it was determined that remediation was needed in the Black Eagle community. The plant grounds and a section of the Missouri River were also designated for remediation, but only after the Black Eagle community was completed. Above is a "Keep off the Flue" sign, measuring 10 by 18 inches, and at right is a metal sample display, measuring 3.75 by 1.75 by 3.5 inches. (Above, 1982.059.1; right, 2007.069.4E.)

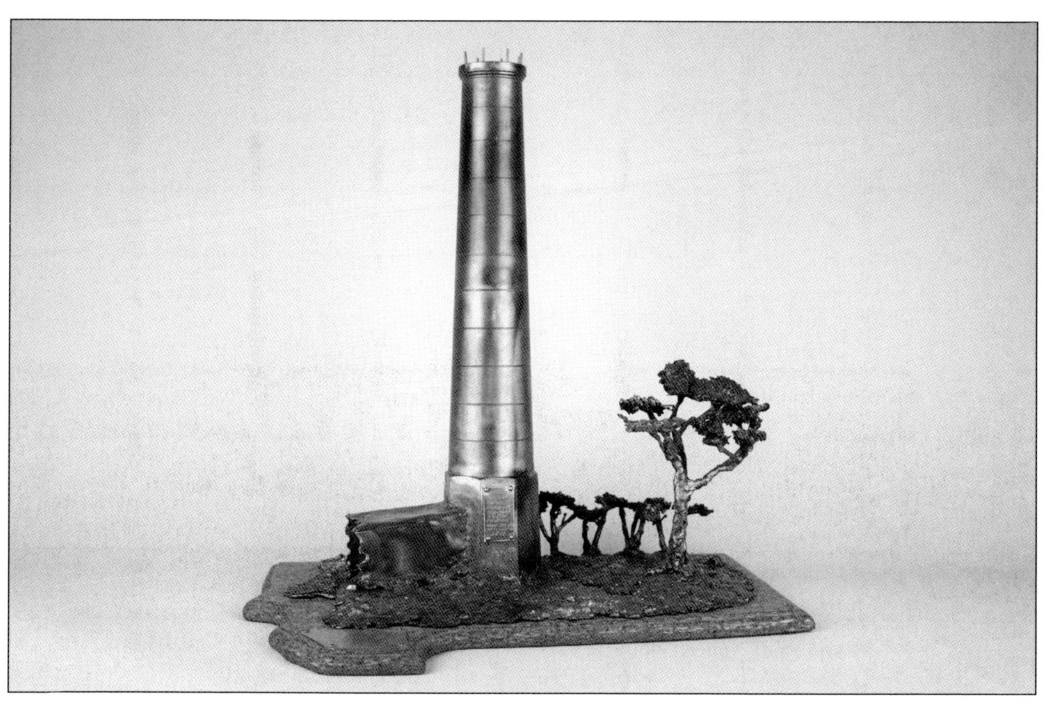

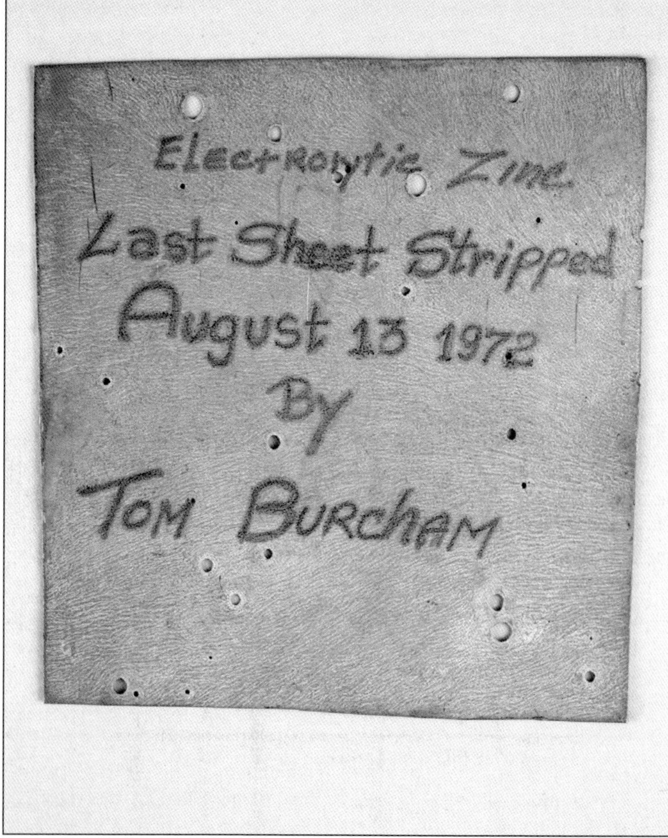

Once the remediation for Black Eagle is finished, the grounds of the plant will be next. There were several spots around the plant where sludge would be temporarily stored in the past and would get into the groundwater. There were also places where arsenic dust was placed due to excessive amounts waiting to be transported. Last, some remediation may need to be done for the Missouri River. From 1893 to 1910, the tailings or waste from the plant went directly into the river. Above is a plaque with copper trees and the Big Stack that was presented to Roy MacRae upon his retirement after 44 years in the machine shop, measuring 15.25 by 9.5 by 14.75 inches, and at left is a piece of the last sheet of zinc from the Zinc Plant, measuring 4.75 by 4.5 inches. (Above, 1999.113.1A; left, 2003.014.1.)

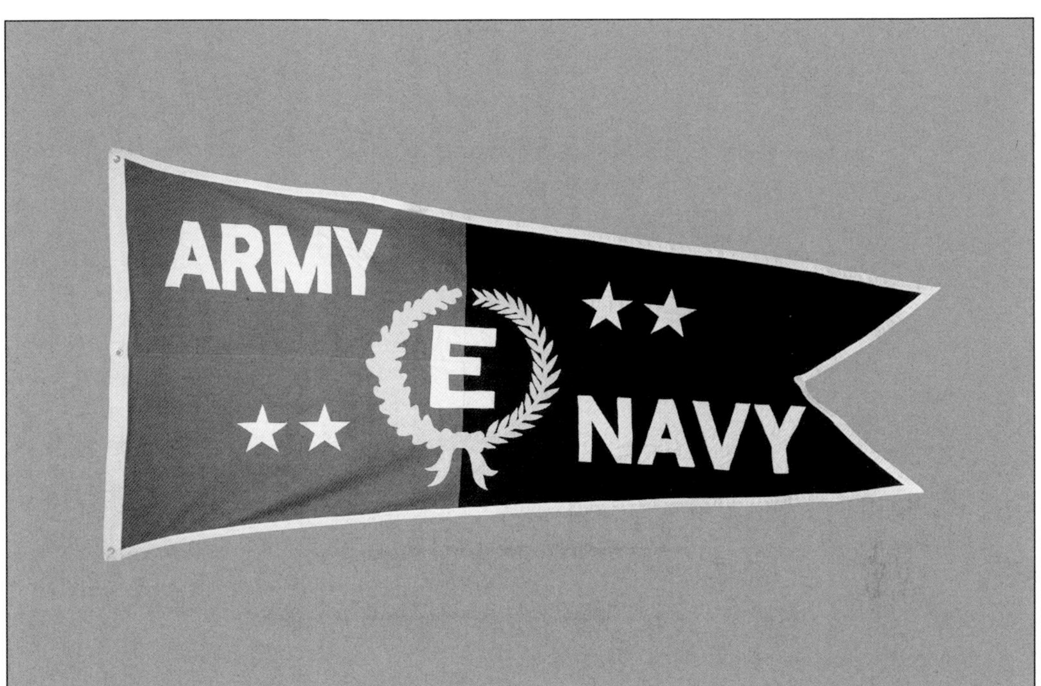

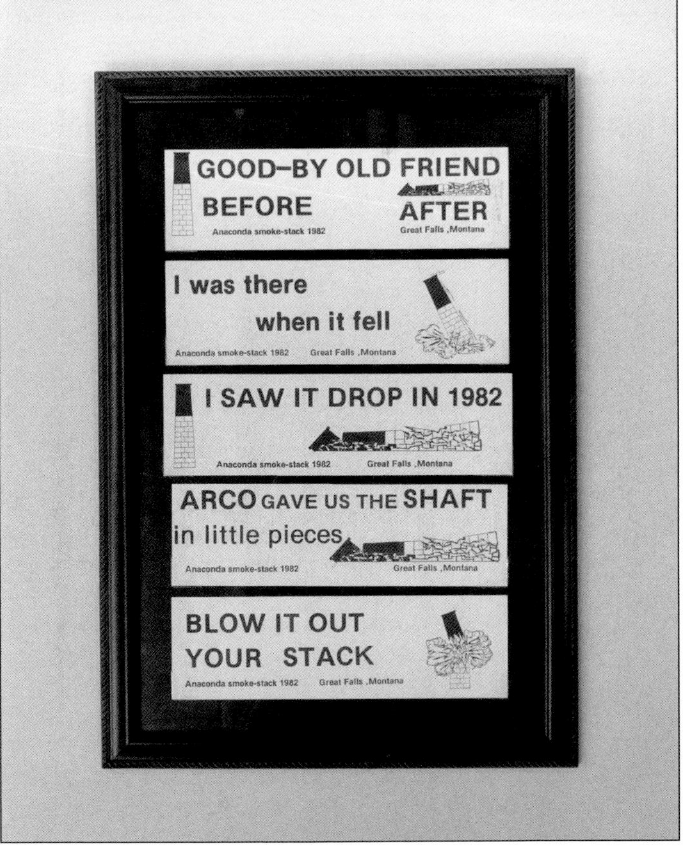

As for preserving the history of the Anaconda Company, many efforts have been made. The Montana State Historical Society did an oral history project studying the metals industry in the state. It interviewed about 25 people from Black Eagle. The Cascade County Historical Society has collected several hundred items and records from the community, putting up displays and producing a documentary for Montana PBS, *Under the Big Stack*. Above is the Army-Navy "E" pennant, measuring 45 by 95.5 inches, and at right is a set of framed "Goodbye Big Stack" bumper stickers. (Above, 1982.017.378; right, 2014.059.1-5.)

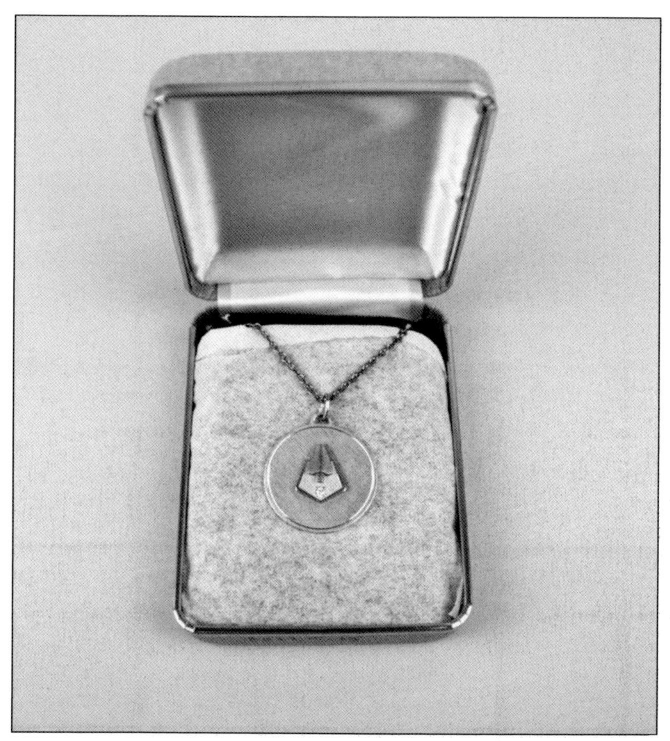

Most recently, another film was produced by the Big Sky Country National Heritage Area, *Spirit of the People: Black Eagle and the Anaconda Company*. The film follows an interview that was done with the last operations director of the Great Falls Plant, Dick Sloan. After his work with ARCO, Sloan went to work with the Montana State Department of Environmental Quality. The Great Falls Reduction Department left more than just the need for environmental cleanup. Pieces of it are still showing up in everyday life, like the necklace given as an employee gift shown at left or the little tools the company used to give out as souvenirs shown below. (Left, 2013.005.15; below, 2014.024.1-10)

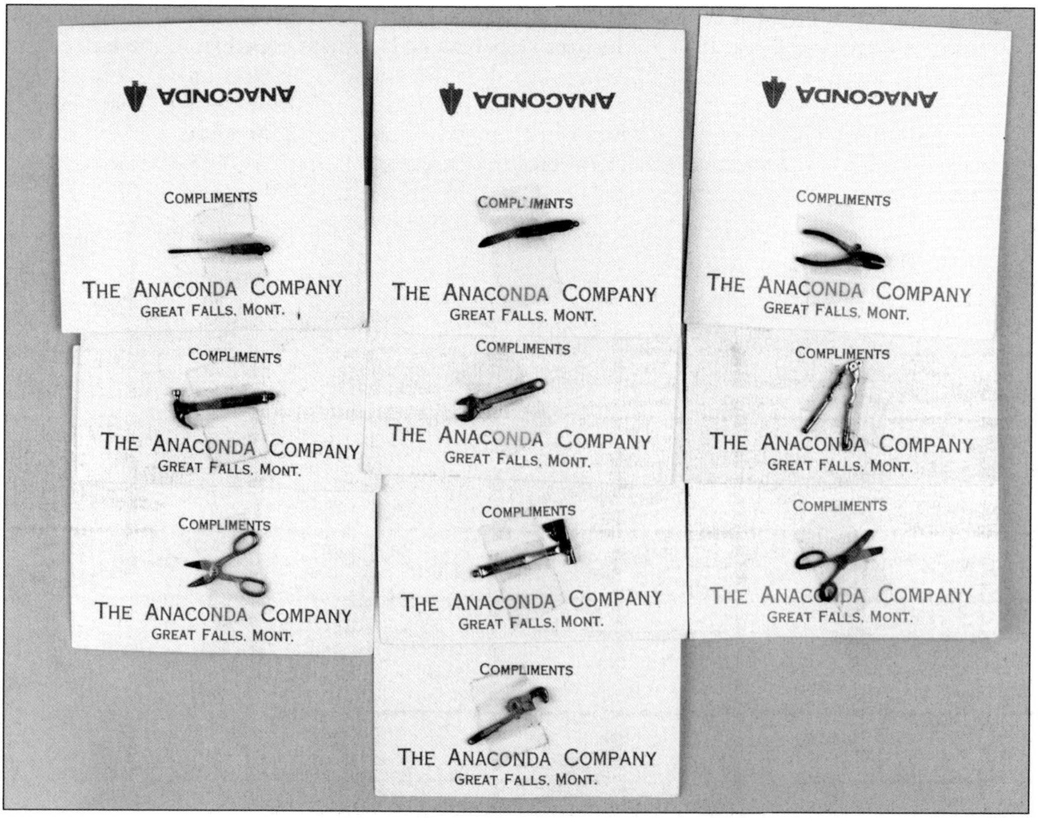

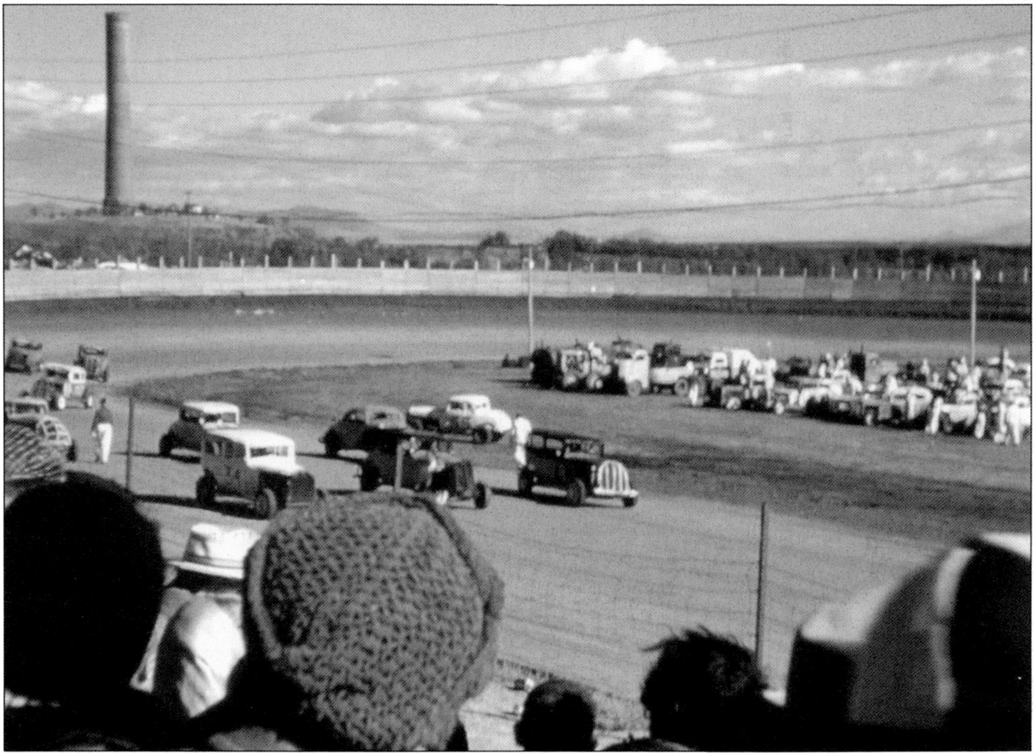

Over 40 years ago, the Great Falls Reduction Department Plant was wiped from the face of the earth. Today, there are still daily reminders of it all around the city, in the names of businesses, paintings by local artists, and even beer named in its memory. It will always be in the background of the community's mind, not physically present but always a part of its consciousness. Above, the Big Stack is visible in the background of a car race in Black Eagle. Below, the Big Stack looms in the distance in a photograph of Giant Springs State Park. (Above, 2020.011.0036; below, 2007.017.0881.)

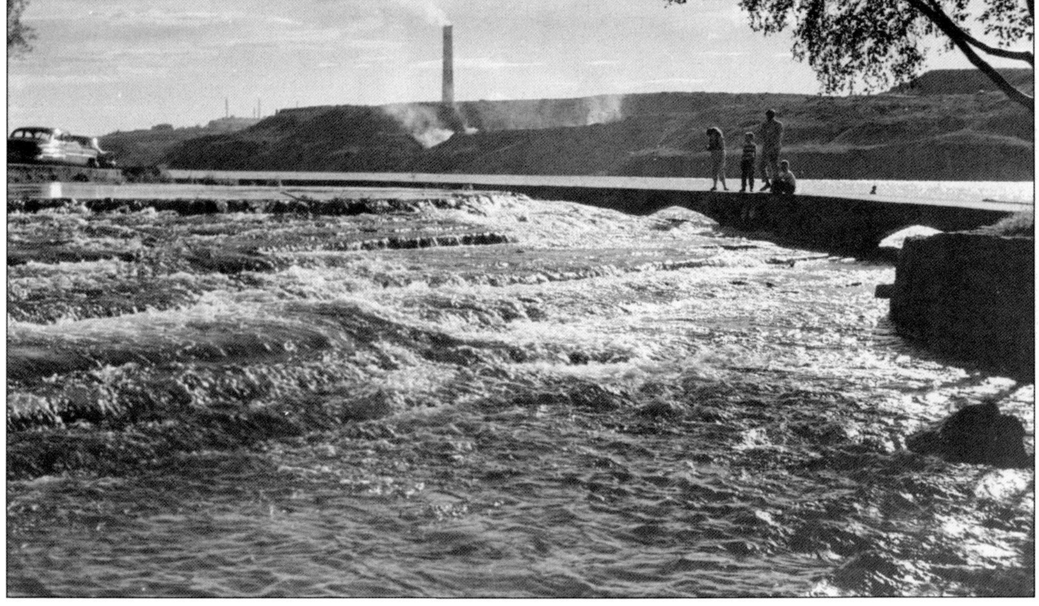

Bibliography

Anode (Anaconda Copper Mining Co.), September 15, 1927.

Arrowhead (Great Falls Reduction Department)

The Big Stack: 1908–1982. Great Falls, MT: *Great Falls Tribune*, 1982.

Black Eagle Book Committee. *In the Shadow of the Big Stack: Black Eagle*. Black Eagle, MT: Black Eagle Book Committee, 2000.

Carl P. Etterer Scrapbooks Binder, 2013.080.1-4. Great Falls, MT: The History Museum & Research Center.

Copper Commando (Anaconda Company)

Gibson, Paris. *The Founding of Great Falls*. Great Falls, MT: Tribune, Printers and Binders, 1914.

Great Falls Leader

Great Falls Tribune

Guy "Tommy" Wever Photo Albums, 2015.089.1 and 2015.089.2. Great Falls, MT: The History Museum & Research Center.

Marcosson, Isaac F. *Anaconda*. Kingsport, TN: Dodd, Meade & Company, 1957.

Oral History Interview 279: Dick Sloan. Great Falls, MT: The History Museum & Research Center.

Special Collection 57: Anaconda Copper Mining Company Papers, 1982.064. Great Falls, MT: The History Museum & Research Center.

Special Collection 69: Anaconda Copper Mining Company Records, 1992.086. Great Falls, MT: The History Museum & Research Center.

Special Collection 560: Frank Bozik ACM Collection, 2022.028. Great Falls, MT: The History Museum & Research Center.

About the Cascade County Historical Society

The Cascade County Historical Society preserves and shares the history of North Central Montana through The History Museum & Research Center archives, collections, exhibits, and public programming. The History Museum Archives maintain 45,500 square feet of closed-stack resources accessible to the public through the Owen & Gayle Robinson Research Center. Historic photographs, blueprints, maps, bound newspapers, yearbooks, government records, and many more original documents are available to the public. The History Museum Collections are comprised of artifacts that illustrate the history of Cascade County and the surrounding Central Montana area. The permanent collection of artifacts includes over 80,000 historic and ethnographic objects for research and use in future exhibitions. Digital cataloging projects are ongoing and will allow collections that are not on display to be available to the public online.

The History Museum & Research Center
422 Second Street South, Great Falls, MT 59405
(406) 452-3462
www.greatfallshistorymuseum.org

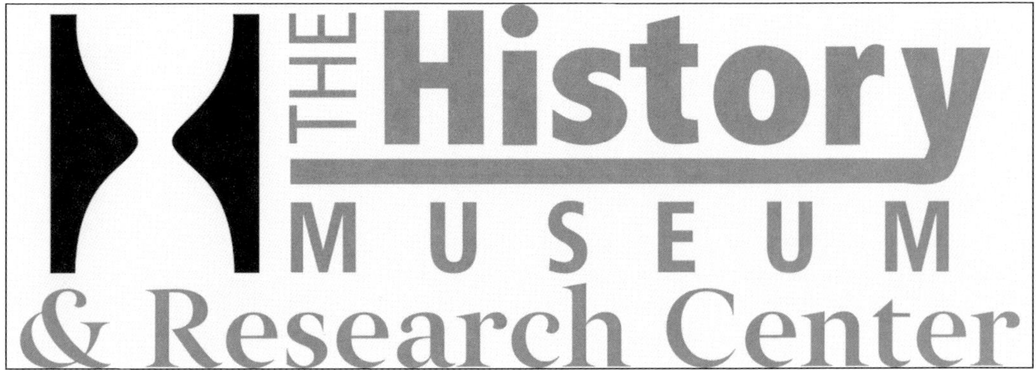

Consistent with our mission to preserve history on a local level, this book was printed in South Carolina on American-made paper and manufactured entirely in the United States. Products carrying the accredited Forest Stewardship Council (FSC) label are printed on 100 percent FSC-certified paper.

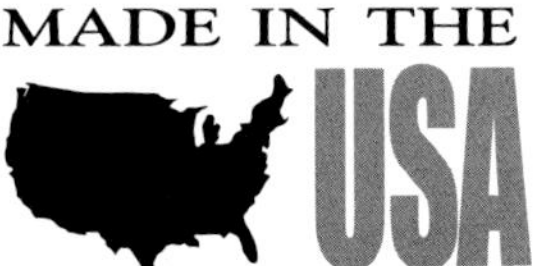